信息收集-形式背后的逻辑

设 计 前 奏

［美］莱斯·沃里克　编著
金雷　译

中国建筑工业出版社

目录
Contents

简介	6	Introduction
设计任务的制定	9	Programming
设计的开始	41	Beginning
场地	69	Place
对话	107	Dialog
文案	129	Document
附录	141	Appendix
词汇表	142	Glossary

致谢
Acknowledgements

　　线和空间事务所关于将信息的收集和整理作为设计工作前奏的理论深受亚利桑那大学建筑学院爱德华·怀特教授的影响，他在教学中一再强调学会在解决问题前发现问题是优秀设计的关键。而怀特教授的这一理论则是受到了著名建筑师威廉·佩纳的影响，他强调通过"寻找问题"的方式去理解客户的需要。

　　感谢本书的版面设计者杰瑞德·罗格。并且还要特别感谢本书的翻译及顾问金雷先生以及线和空间的设计团队成员鲍勃·克莱门特先生、亨利·汤姆先生、约翰·博金班先生、麦克·安格林先生、艾米丽·斯塔斯女士、约翰·麦考恩先生以及曼尼·克洛普夫先生，他们的设计实践为本书提供了丰富的实例。另外还要感谢苏珊·沃里克女士对本书富有见地的评论。书中的照片由罗伯特·莱克、比尔·蒂默曼、维尔特、麦克·托雷、莱斯·沃里克、鲍勃·克莱门特、麦克·安格林、杰瑞德·罗格和亨利·汤姆提供。

The Line and Space method of becoming informed as a precursor to design was heavily influenced, by Edward T. White, Professor at the University of Arizona College of Architecture, who taught that identifying problems before solving them was fundamental to design excellence. Professor White, in turn, was affected by the famed, Architect, William M. Peña and his "Problem Seeking" approach to understanding client's needs.

Thank you to Jared Logue our book designer, illustrator and editor. Thanks, also, to our translator and advisor, Lei Jin and the Line and Space team, Bob Clements, AIA, Henry Tom, AIA, Johnny Birkinbine, AIA, Michael Anglin, AIA, Emily Starace, RA, John McColgin and Manny Kropf, RA, who contributed greatly to the examples shown in this book and thanks to Susan Wallach for insightful commentary. Photography was by Robert Reck, Bill Timmerman, A. T. Willett, Mike Torrey, Les Wallach, Bob Clements, Michael Anglin, Jared Logue, and Henry Tom.

简介
Introduction

"信息收集–形式背后的逻辑"作为本书的标题，将本书的想法浓缩进一个组合词里，那就是说建筑设计、信息的收集整理以及最终呈现的形式是密不可分的整体。

这是一本有关如何制定设计任务的书，即试图去描述如何获得对于建造项目实际需要的一个清晰理解的过程。作为建筑师，在真正的设计工作开始之前，在设计任务的制定中，我们试图去寻找各种信息，可以将建筑设计的过程转化为解决问题的过程。这些信息可以被看作是片段的拼图，当问题解决后，就会将建筑师的哲学、直觉和情感同文化元素、社会学、经济学、环境当然还有人们的需要融入最终的建筑里。

我们会通过线和空间建筑师事务所的一些实际案例来说明如何定义问题，由成功项目中选取的任务书、图纸或是照片组成的插图进一步帮助强调出信息收集的价值。材料被分为五个章节，后面辅之以术语列表，以及一个包含有案例分析的附录。第一个章节，设计任务的制定，讨论制定可以成为创造成功建筑的保障性文案的必要性。本书接着会带领读者了解在制作设计任务文案之前所必须完成的三步工作：第一步，开始，在制定设计任务之前搞清楚需要完成的任务；下一步则是解释为什么场地是一个设计项目的主要决定因素；三步中的最后一步，对话，告诉读者如何从各个方面去定义出客户的需求。最后，以文案作为标题的章节，以实际的案例分析为基础，提供了一些组织和表达从设计任务的制定过程中收集到的信息的不同选择。

我们希望建筑师、设计师、业主、代理商，以及各个层级的客户都可以通过本书深刻了解线和空间事务所关于将信息的收集整理作为设计工作前奏的理论。《Informed》将会为如何设计出以客户需要为中心、以技术知识为基础、符合时代及场地要求的生态建筑提供有用的技巧。而这一理论也支撑了线和空间事务所超过35年的设计实践。

---

*in***formed**, the title, embodies the idea imbedded within a single word, that in building design, information and form are inseparable.

This book is about *Architectural Programming*; the process of gaining a clear understanding of building requirements. As architects, before beginning to design, through the act of programming, we are participating in a search for knowledge that will help us define creating the building as solving a problem. This information can be seen as the pieces of a puzzle which when solved, weaves the architect's philosophy, intuition and emotion together with elements of culture, sociology, economics, environment, and of course, peoples' goals, needs and wants into architecture.

The idea of problem definition is supported with a number of specific examples drawn from the work of the architecture firm Line and Space, llc. Illustrations using actual programs, drawings and photographs of successful projects help reinforce the value of being informed. Material is organized into five chapters followed by an appendix including case studies and a glossary defining terms. The first chapter, **Programming**, discusses the need for developing a document that becomes the umbrella for creating successful architecture. *in*formed then takes the reader through three steps that must be completed in advance of creating the program document. The first, **beginning**, makes clear the things to accomplish before programming can start. Next is an explanation of why **place** is a primary determinant for design of a project. The last of the three steps, **dialog**, takes the reader through all aspects of defining client requirements. Finally, the chapter titled **document** offers options, based upon case studies, for organizing and presenting the information gathered during the programming process.

It is hoped that architects, designers, owners, agencies, institutions and clients of all levels of sophistication gain insight into the Line and Space method of becoming informed as a prelude to design. *in*formed will provide useful techniques for creating a client focused, knowledge based, ecological architecture of its time and place. At Line and Space, llc for over 35 years, knowledge has illuminated design.

设计任务的制定
Programming

知识同形式是密不可分的。在建筑师开始设计创作之前,他们必须要了解什么是一个项目的核心要素。

Knowledge and form are inseparable. Before architects begin to create they must understand the essential elements of a project.

知识同形式是密不可分的。

在建筑师开始设计创作之前，他们必须要了解什么是一个项目的核心要素。建筑设计任务的制定，一个发现并记录下这些基本要素的过程，由两个基本的方面组成：场所，即建筑所处的场地及它所带有的其他所有信息（文脉、文化、地形、气候等）；对话，即以建议、目标或是事实陈述等各种不同形式所表现出的客户的需求。这两个同等重要有时甚至是互相矛盾的要素决定了建筑应该是什么样的。更简单地说，引用一句建筑教育中的经典名言："形式追随知识（功能）。"

建筑设计是关于解决问题的。但什么是问题呢？为了找出问题，建筑师会开始一个研究、质询和观察的过程。使用者在这个过程中会参与进来。他们会被问：你们是谁？你们是干什么的？你们是怎么工作的？你们需要什么？你们如何学习、教学和互动？以及更多的问题：离某些人或者地方近一点会提升你所做的事情吗？你是怎么到达的？你们的人员构成是怎样的？你需要和谁打交道？是群体还是个体，成年人还是儿童？这些人是谁？事情进展的如何？要花多少钱？为什么？什么时间？接下来如果有一个场地，那么微风从什么方向吹过来？是季节性的吗？夏天和冬天的太阳高度角有什么不同？最热的时候有多热？最冷的时候呢？地质条件如何？水文条件又是怎么样的？全球变暖的问题有考虑过吗？谁之前住在这儿？我们从他们身上能学到什么吗？颜色？质询的同时，问题被逐渐揭示出来。

Knowledge and form are inseparable.

Before architects begin to create they must understand the essential elements of a project. Architectural Programming, the process of discovering and documenting these fundamentals, is informed by two primary factors, **place**; the project site, along with all its forces (context, culture, terrain, climate, etc.) and **dialog**; the client's needs in the form of precepts, goals and facts. These two equally important and sometimes competing elements guide what the architecture shall be. More simply put and restating the teaching of an architecture master: *form follows knowledge*.

Architecture is about problem solving. But what is the problem? To find out, architects begin a process of research, inquiry and observation. Stakeholders are engaged. They are asked: Who are you? What do you do; how do you do it? What do you need? How do you learn, teach, and interact? And more questions: Is what you do enhanced by proximity to someone or someplace? How do you arrive? Who are your constituents? Do you deal with crowds or individuals; adults or children – who are these people? What's going on? How much? Why, when? And then there is place. Where does the breeze come from? Is it seasonal? What is the differential between winter and summer sun angles? How hot does it get? How cold? What is the geology? The hydrology? What about global warming? Who dwelled here before? What can we learn from them? Color? Questions are asked – the problem is revealed.

设计任务的制定
Programming

建筑设计任务制定的过程让建筑师有机会提出问题、创造对话，从而为指导设计提供深度的有价值的信息。

对于从对话过程中收集的信息及问题的回应被仔细记录在案。

The process of architectural programming allows the architect to ask questions, create dialog, and gain insight that will provide an invaluable resource that guides design.

Responses to questions and information gathered from dialog is carefully documented.

设计任务文案，是一个对于所有定义出的建筑问题信息的汇编，实际上就成为我们的设计手册。当"设计如何去满足项目的要求？"这样的问题出现时，这本手册将会同时为建筑师和业主提供答案。设计任务的制定可以保证无论最终的建筑以何种形式呈现，它都会去真实地满足项目的要求。在设计讨论的初期，有一点必须非常明确，即驱动设计的是预算、生态、功能需要——而不是风格或是时尚潮流。

我们经常会发现一些似乎是善意而为的设计其实并不能称之为"建筑"。更进一步讲，这些项目的设计是在没有提出足够多的问题、没有理解什么是对场地和客户最重要的因素的基础上做出的。诚然，作为一个设计师，很难抑制在有一个简单的概念前就去进行创作的冲动。但是，在我们收集到足够多的信息之前，这种冲动所带来的成果其实没什么实际意义。通常来说，对于一个设计师而言最难的事情是遵从这样的准则，即等到问题被清晰地定义出来之后再开始去寻找解决方案。

总而言之，在和建筑师沟通交流一个建筑项目的需求时，业主会遵循下面的四种方法。

按照详细程度来分，它们是：①纲要性的设计任务——在这个情景里业主向建筑师提供一份最精简的信息，期待着一个标志性的形式或是商业化的形象，而不去强调建筑的实际功能如何运作或是如何融入到场地中；②标准设计——在这里，业主会提供给建筑师适应当地要求的平面和立面，只需要建筑师进行一些必要的调整以满足当地的条例规范要求；③业主提供的设计任务

The program document, the compilation of all of the information defining the architectural problem, becomes, in fact, the *Design Manual*. It is here both the client and the architect go when question arise over how the design is meeting project requirements. The program provides assurance that no matter the form the architecture takes it is true to project aspirations. In early discussions, it must be made clear that budget, ecology, goals, needs, and wants drive the design —— not style or fashion.

It is not uncommon to find seemingly well-intentioned buildings that are simply not Architecture. Most assuredly, the project was designed without asking enough questions; without understanding the issues important to the site and the client. Admittedly, as a designer, before one has little more than mere notions, the temptation is to create. But until we are informed, the outcomes of those inclinations have little real meaning. Often, the hardest thing for a designer to do is to have the discipline to wait until the problem is defined to begin work on the solution.

Generally, to communicate the requirements of a building project to their architect, owners follow one of **four methodologies**.

These are, in order of detail: (1) **Outline Program -** In this scenario the owner provides a minimum amount of information to the architect, expecting an iconic form or marketable image with almost no emphasis on how the building actually functions or engages its site; (2) **Standard Design -** Here, the client provides the architect with floor plans and elevations to be adapted to a specific locality requesting only changes necessary to satisfy local codes and ordinances; (3) **Owner Provided Program -** This methodology, often preferred by institutions with

书——这种方法通常适用于配有相应管理部门的机构，他们会提供给建筑师一份由内部工作人员或是外部咨询机构所编纂的设计任务文案。在这种情况下，一个新被挑选的设计师会被任命去审核并且确认或者扩充其中的内容。当然，最终最能够平衡好场地环境和客户利益之间关系的创造建筑的方法是：④建筑师主导制定出设计任务——为了保证定义问题和解决问题过程的连续性，建筑师主导制定设计任务的方法受到下列因素的影响，即所雇佣的建筑师事务所是否最符合客户的理念并且能够积极参与到客户制定项目发展计划的过程中。

纲要性的设计任务

在有些时候，建筑师会被要求在已知很少甚至是根本就没有设计任务信息的情况下设计一个建筑。在这种情况下是否可以创作出有意义的建筑作品是非常值得怀疑的事情。一般情况下，这样的压力通常来自开发机构或是团体，他们没有资金，但是需要靠一些"形象"来销售他们的项目或是筹募资金。对于前面一种情况而言，如果是开发商驱动的项目，那么我们对于未来空间实际的使用者的需求所知不多。我们会甄选出典型的使用者并赋予建筑一些一般性的功能定位，例如专卖店、办公室、工业厂房或者是综合体。这样的设计工作最终也可以转化为真正的建筑。为了这个目的，开发商必须要提供这种建筑类型中真实的功能和组织架构。另外，同其他所有的优秀建筑设计一样，建筑师需要充分地理解并恰当地回应建筑所处的场地。

---

facility departments, provides the architect with a program document prepared by either in house resources or an outside consultant. In this case, a newly selected designer may be commissioned to review and confirm or expand its contents; and, finally, the most preferred methodology for creating buildings that best balance environmental and client's needs: (4) the **Architect Directed Program -** Assuring continuity between problem defining and problem solving, the architect directed program is effected by hiring the architecture firm most compatible with the client's philosophy and then participating with them as they facilitate the development of the building program.

Outline Program

There are situations when the architect may be asked to design a building with little or no programmatic information. This is a precarious situation for a meaningful architectural outcome. Most commonly, the pressure to design in this manner will come from the development community or groups who simply have no money but need an "image" to sell their project or raise funds. In the former case, if developer driven, it may be that little is known about the actual needs of the future space occupier. Generic tenants are identified and the building is deemed to be a general use such as retail, office, industrial or some combination of these. This type of work can become real architecture. In order for this to occur the developer must provide information on the functional and organizational realities of the building type. In addition, as with all good architecture, the architect must fully understand and respond to the site.

如果一个客户，例如一个集团的总监或是一个机构想要建筑师为他们提供一个"漂亮的建筑景象"来作为去筹募资金的工具，那么这实际上是要求建筑师为他们创造出一个假象。具有讽刺意味的是，这个看起来合乎逻辑的要求其实是对于建筑创作的误导。在这样的情况下建筑师可能会解释说他们的手上没有足够的信息来确定建筑应该长成什么样子。而客户，则因为急需要筹募到资金，会告诉建筑师这些建筑方案只会被用来诱使潜在的投资人为项目投资，当资金到位后，建筑师就可以着手去设计真正的建筑。这种方法的问题在于，除去要设计一个虚假的方案，投资者有可能并不接受一个与当初他们看到的建筑方案不一样的设计。

根据我们的经验，很少有投资人（占总数的20%或者更少）的投资额会占到一个新建项目所需投资总额的大多数（例如80%）。这些资助人会非常希望看到一份完备的设计任务计划，因为这就表明这个项目有一个富有见地的领导者愿意去细致地发掘出他们的实际需求，而这也反过来说明项目的预算是基于一个现实的情况来做出的。一份经过深思熟虑而制定出的建筑任务计划会给予潜在的投资者以信心，它表明项目的领导者已经做足了功课。建筑任务计划本身在筹募资金的过程中会成为唯一必要的工具。

在主要投资人真的非常关心未来建筑外观的情况下，开发机构在处理同他们的关系时存在着风险，因为向他们展示的建筑外观可能会是一个错误的"景象"。于是建筑设计变成了一个二次猜测投资人喜好的过程，建筑创作是去创造出一个假象而不是真的去考虑如何满足客户的需求。

If a client such as the director of a group or institution desires a "pretty picture" to use as a fund raising tool, it is really asking the architect to create an untruth on their behalf. Ironically, this seemingly logical request is so misguided. What happens in this situation is that the architect explains that he does not have sufficient information to know what the building should look like. The client, desperate to raise money, states that this pretty picture will only be used for enticing potential donors to help fund the project and after money is raised, the real building can be designed. The problem with this approach, besides dealing in untruth is that donors may not accept a building different from the image that they have already seen.

In our experience, a very small number of donors (on the order of 20% or less) contribute the majority (say, 80%) of the funds required for a new building. These patrons are interested in reviewing a well-conceived building program because this indicates that the project has insightful leaders willing to carefully define the group's needs and wants which in turn means that budgeting is based upon realistic scenarios. A thoughtfully created Architectural Program gives potential donor's confidence that the project leadership has done their homework. The program itself becomes the only fundraising tool necessary for seeking major gifts.

In the case where a major donor really is concerned with what the architecture looks like, the group or institution places their relationship with this potential contributor at risk by possibly showing the wrong "picture". Making architecture then becomes a process of second-guessing what the donor might like and creating a sort of fantasy rather than fulfilling the client's needs.

设计任务的制定
Programming

通过向客户（下方图片中所示）传达一份详细的场地分析（上方图片中所示）和确认的设计任务书，我们一致决定项目首要的任务就是在尽可能地减少对野生动物产生影响的前提下，将项目融入沼泽栖息地中去。在收集到这一重要信息之后，我们绘制了右侧的概念示意图。图中显示了新的建筑（在图中橙色的部分）将会建造在沼泽边缘之前被干扰过的场地上，场地提供了看向栖息地方向无与伦比的景观，同时也最大限度地减少了对于栖息地内现存生物的影响。

By conducting a detailed site analysis (image above) and program confirmation with the client (image below) it was determined that the ability to be integrated within the marsh habitat while minimizing impact to wildlife were the top priorities. After gathering this important information, the concept diagram to the right was drawn, illustrating how the new facility (orange in diagram) sits on previously disturbed land at the edge of the marsh, providing unparalleled views to the refuge habitat and minimizing impact to existing native habitats.

最后，还有一种比较少见的情况，业主热衷于造出一个地标，而对于成功的功能设定则没有什么概念。在这种情况下，建筑师有责任反向根据惯例提出一份设计任务。另外重申一遍一条最基本的原则：花时间真正地去理解建筑所在的场地将会引领设计师通向成功。

标准设计

零售商或是一些机构客户向建筑师提供一些带有需要紧密遵循的设计方向的标准设计并不是一件多么不寻常的事。零售商希望他的顾客能够熟悉他们以前已经建立起来的标准布局，这样不管顾客去哪家分店购物，卷心菜、锤子或是灯泡总可以在同一个地方被找到。理智告诉我们这种对于布局的熟悉可以节约顾客的时间并帮助销售。可能这么做确实帮助到了销售，但是有没有考虑到顾客的人员构成呢？环境呢？购物的体验又是如何的呢？

Finally, in the rare case when the owner is most interested in an iconic form and has little actual experience with the functions required for success, the architect has a responsibility to reverse engineer a program from an often sketchy list of requirements. Once again, at a minimum, spending the time to gain a real understanding of the environment in which the building will sit will guide the designer to success.

Standard Designs

It is not unusual for retail and even institutional clients to provide **standard designs** to their architects with the direction that these should be adhered to closely. In retailing the owner wants customers to be familiar with their establishments' typical layout so no matter which store in a chain is visited, the cabbages or hammers or light bulbs are always in the same place. Logic tells us that this familiarity saves the customers time and facilitates sales. Possibly. All fine and good for sales, but what about the demographic? The environment? The experience of shopping?

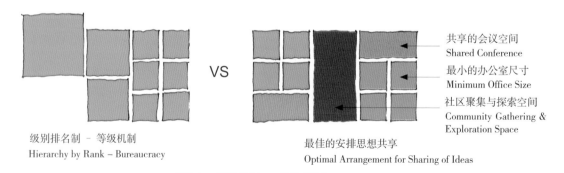

图1.1 等级机制 vs 协作机制
Figure 1.1 Bureaucracy VS Collaboration

当我们可以很高兴地确定全麦面包在什么位置的时候，难道我们不渴望变化、惊喜、探索和一定的神秘感吗？与其去担心"不易寻找"，不如去尝试一种"有趣的发现"和探索，这样可以带来一个好的建筑设计并随之提升销售业绩。

有相似的情况，一个政府机构可能会发展出一套能反映出他们雇员服务等级（基于教育和经验层次的分级）的办公空间设计。一般来讲，一个雇员的级别越高，他就可以获得与之级别相匹配的更大的空间和更多的设施。当工作职责被清晰限定并且功能要求拥有更大的空间意味着负担更多责任的时候，这么做是有意义的。但是，这种标准化的设计变成了一种身份标签，高级别的员工不愿意自己的办公空间比低级别的同事小或是装备比低级别的同事差，哪怕他们的工作实际只需要一张桌子就可以完成（图1.1）。

这样看来，将办公空间的大小同员工级别直接挂钩的做法会阻碍创新。这实际上指出了使用标准化设计的弊端。举例来说，一个有着一间大办公室但是需要经常合作和冥思的人，有可能最适合他的就是一个很小的空间，旁边配有智能化会议室，一些鼓励谈话交流的角落和可以安静进行创造性思维的花园。

线和空间建筑师事务所曾经被一家政府机构任命依托一个标准设计来做一个项目以期节约时间和预算。在和客户进行的设计任务研讨会上，我们将这个标准设计当作是一个前提条件而不是

While we are happy to know with certainty where the whole grain wheat bread is, don't we crave variety and a sense of mystery, surprise and exploration? Instead of worrying about "hard to find", try: "fun to find" and the exploration could lead to good architecture along with increased sales.

In a similar manner, a government agency may have developed a standard staff office design reflecting their employee's Service rating (hierarchal rating based upon education and experience). Generally speaking, the higher one's rating, the more space and amenities that person is entitled to. This makes sense when job descriptions are examined and functions dictate extra space to accommodate increased responsibilities. However, this standard soon becomes a marker for status and a higher ranked individual is disinclined to have a smaller, less well equipped, space than a less senior co-worker even though job performance may not require much more than a desk (Figure 1.1).

Taken literally, the prescription of office size related to rating will preclude innovation. This reality points out the problem with using standard designs. For example, a person entitled to a large office but having the need for collaboration and contemplation may be best served by using a very small space supported by smart conference rooms, nooks and nodes for encouraging discourse and gardens for creative thought.

Line and Space was commissioned by a government group that hoped to rely on a standard design to facilitate time and budget savings. Programming sessions with the client utilizing this design as input, not dogma, and thoughtful discussions regarding actual requirements of the building, led to the realization that

必须遵守的教条，之后针对建筑的实际需要作深入讨论，让大家意识到真正的问题不是按照他们习惯的方式"怎么去布置办公室"，而是"如何在一个敏感的场地中布置建筑。"在随后几天同使用者的对话中，更进一步确立了目标，包括让员工可以保持和周边环境的紧密接触、增进和谐沟通交流的能力，以及最重要的一点，保护一种生活在场地中的濒危鸟类：长嘴秧鸡。

---

the real problem was not "How to arrange offices" as illustrated in their customary plan but rather; "How does one best situate a building in a sensitive habitat?" Further goals developed during several days of dialog with the stakeholders included enabling staff to maintain a strong connection to their surroundings, encouraging the ability to engage in harmonious communication and most importantly, protection of an endangered bird located on site: the Clapper Rail.

从北面看向圣地亚哥国家野生动物救助综合体，照片中展示出的倾斜的玻璃幕墙为建筑与环境景观建立起了强大的联系，同时由于反射的是地面而不是天空，大大降低了飞鸟误撞的可能性。

The San Diego National Wildlife Refuge Complex from the North, showing angled glass walls that provide a strong connection to the landscape while mitigating bird strikes by reflecting the ground rather than the sky.

设计任务的制定
Programming 19

The lesson here is that the architectural problem (or what the building should be) is not always obvious.

这张由英文单词构成的平面构成展示了建筑师参与定义问题过程的重要性。经常出现的情况是，第一眼看过去似乎是显而易见的问题在有针对性的调查研究完成后才发现最初的理解并不完善。

This word art graphic illustrates the importance of architect facilitated problem definition. Often, what seems like an obvious problem at first glance, is not fully understood until focused study and investigation are complete.

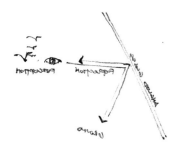

每年都有数以百万计的鸟类因为撞击玻璃窗而死亡，因为它们会将反射在玻璃里的天空误认为是它们的飞行路径。而另一方面，在办公室或是实验室中工作的人，尤其是进行生态环境方面研究的科学家，渴望同室外环境之间有直接的视觉联系。而可以观察周边环境最好的方式就是通过使用大面积的落地玻璃幕墙来实现。当这些看似矛盾的元素被发掘出来后，问题就变为了如何将落地玻璃幕墙最大化以增加员工同外边环境间的视觉联系，同时又尽可能地减少鸟类撞击玻璃事件的发生。相关的研究发现当大面积的玻璃幕墙面呈现出一定的倾斜角时会产生这样一种情况，即飞到附近的鸟类从玻璃的反射中看到的不是天空，而是地面（图1.2）。因此，为了避免撞击到它们观察到的所谓陆地，长嘴秧鸡会选择飞离玻璃而不是撞上去。当知道这样的信息后，客户承认他们原先的标准设计并不能解决手边的实际问题。由于同意让知识引领设计，客户最终收获了满意的建筑，既保护了鸟类，又增进了员工间的互动以及室内室外之间的联系。

这个例子说明了建筑设计所需要解决的问题（或者，建筑物应该是怎样的）并不总是显而易见的，标准设计不会自动满足客户的功能或者是环境方面的真实需求。

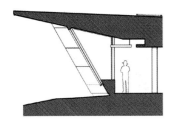

（图1.2）一系列的示意图展示了倾斜的玻璃幕墙如何为室内的员工与室外的自然环境之间建立起来强大的联系，而同时保护飞鸟不与玻璃幕墙发生误撞。
(Figure 1.2) The sequence of diagrams above illustrates how the angling of glass walls provides staff with a strong connection to the outdoors while protecting birds from colliding with the building.

Every year millions of birds die by crashing into windows since, as they approach a building, they perceive the sky reflected in the glass as part of their flight path. On the other hand, people working in an office or laboratory, especially scientists engaged in environmental study, crave a visual connection to the outdoors. This ability to observe their surroundings is best obtained through the use of large expanses of floor to ceiling glass. Once these seemingly contradictory issues were identified, the ability to maximize glass to connect staff to the outside while at the same time minimizing it to reduce bird crashes the problem became defined. Research led to studies which showed that the large expanse of glass desired, could, by simple angling, create a condition so that as birds flew in the vicinity they would view a reflection of the ground, not the sky (Figure 1.2). Hence, to avoid crashing into what they perceived as terra firma, the Clapper Rail would fly away from the glass and not into it. Once this information was shared, the client agreed that their standard design did not solve the problem at hand. Letting knowledge guide the design, the agency received a building that protected birds, encouraged staff interaction and connected those inside with the outside.

The lesson here is that the architectural problem (or, what the building should be) is not always obvious and a standard design is not going to meet the client's functional or environmental needs.

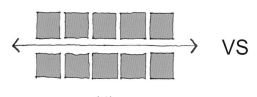
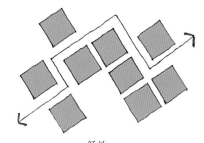

（图1.3）示意图的比较展示出加权系数是如何影响到设计的创造性的。
(Figure 1.3) Diagram comparison showing how grossing factors can affect the creativity of a design.

业主提供的设计任务书

通常，一些常规基础上的组织，例如大学或者是公共学校会将编写建筑功能作为他们工作计划过程的一部分。这个工作可能是交给内部的工作人员，也可能外包给外面合适的专家去做。尽管内部计划的编订人员或是外聘的咨询专家可以利用他们的专业知识来确定预算或是选定场地，但一般来讲这些设计任务的深度都不足以帮助建筑师深刻理解建造场地或是明确建筑应该是怎样的，建筑师通常只能从这些信息中了解建筑的大概类型，例如图书馆、实验室或是教室。

通常来说，当决定空间分配的时候，设计任务的制定者，因为缺乏时间或者意愿，经常会犯两种错误：一种是设计任务的制定者使用公开发表的或者是内部通用的标准，而不是依据实际的功能活动

Owner Provided Program

Often, owners who build on a regular basis such as Universities or Public School Systems, will write building programs as part of their planning process. This may be done in house or contracted out to appropriate experts. While the in house planners or consultant may have been able to use their programming skills to establish a budget or select a site, these programs do not generally go into the depth necessary for the architect to understand the building location and what the building should be beyond a generic typology, say, for example, library, laboratory or classrooms.

Often, when determining space allocations, the program creator, due to lack of time or inclination, makes two consequential errors. One, the programmer utilizes published or internal standards instead of calculating the space requirements based upon anticipated function, movement and actual equipment and furnishing sizes. Secondly, a number known as the grossing factor which adds space for circulation, structure, exterior walls, interior partitions, HVAC equipment is insufficient or omitted entirely. The grossing factor, varies according to building and space type and whether support spaces such as storage, mechanical rooms, etc.

以及设备家具的尺寸来计算出所需的空间面积；二是，一个被称为补偿系数的面积数量经常会考虑不够或是直接被忽视掉，这包括了交通空间、结构、外墙、室内隔断、暖通设施等所带来的面积增加。补偿系数会根据建筑以及空间的类型而变化，另外还要考虑一些支持性的空间，例如储藏、机电用房等是否在面积分配的时候已经被包括在内了。根据我们的经验，底层建筑这个数值一般在20% ~ 35%。高层建筑则会效率更低，引用SEV和ÖZGEN的数值，在30% ~ 40%（S Aysin, ÖZGEN Aydan. Space: Efficiency in High Rise Office buildings [J]. Metu Journal of the Faculty of Architecture,2009, 26(2):Table3.)（图1.3）。

---

have been included in the space allocation. In our experience, this number averages between 20%~35% for low structures. High rise buildings tend to be less efficient and as SEV and ÖZGEN cite, run 30%~40%. (S Aysin,ÖZGEN Aydan. Space:Efficiency in High Rise Office buildings [J]. Metu Journal of the Faculty of Architecture,2009,26(2):Table3.) (Figure 1.3)

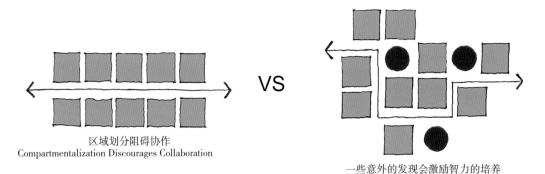

图1.4　最优化的效率 vs 鼓励协作
Figure 1.4　Optimal Efficiency VS. Encouraging Collaboration

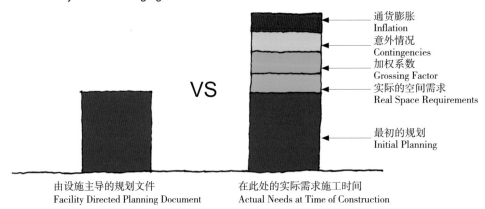

图1.5　创建真实的预算
Figure 1.5　Establishing True Budget

那些以最大化"建筑效率"自居的设计任务书的制定者一般都是以牺牲建筑设计的创造性为代价的,造成的结果就是建筑只剩下最小宽度的冗长的通道,紧挨走廊两侧布置的房间以及毫无特色的门厅。理性的分析可以得出这样的结论,例如,在一个学术环境中,这样的建筑抑制了非计划内空间的创造。最大化的空间效率消除了很多可能性,例如可以增加合作精神的衬着写字板的超宽走廊或是角落空间。渴望交流合作与提升空间效率之间的矛盾可以通过简单地增加5%~10%的补偿系数来解决(图1.4)。

由于设计任务的编纂者一般也是财政预算的编订人,这些人通常都会设置这样一个陷阱,即项目可用的预算是基于一个极度节省的平面布置的前提下做出的。在资金不足的前提下被要求设计一个能够满足使用者需求的设施是很常见的事情。而这种情况还可能会因为预算准备期同实际施工期之间施工费用的上涨而进一步恶化。在对于业主来讲最好的时间里,材料及人工费用的上涨都可以保持在一个较低的比率,然而在施工项目繁荣发展时带来的价格上涨可能会是毁灭性的。例如,在2013年,一个相对低迷的市场,工程新闻记录杂志报道显示施工价格水平的年增长率保持在2.5%以下。如果预算的制定者没有意识到在最初的项目前期讨论与最后的建造实施期间最少4年的时间差,也许项目预算的10%就会以某种形式在某个地方被拿掉,从而导致项目规模的缩减或是工程质量的下降。并不是很久以前,在2007年,拉斯维加斯的建设综合成本以每月1.5%的幅度增长。这换算成年增长率就是20%。在这种情况下,显而易见,经过4年的时间,项目的预算需要翻倍,以应对快速增长的建造成本和突发事件(图1.5)。

Programmers who pride themselves in maximizing "building efficiency" generally stymie creativity by not allowing for buildings to have anything but endless hallways with minimum width, double loaded circulation and anonymous doorways. When taken to its logical conclusion, for example, in an academic environment, this type of architecture, inhibits the creation of unprogrammed space. Maximum efficiency eliminates possibilities such as extra wide corridors or niches lined with whiteboards, that promise a certain serendipity tending to incubate collaboration (Figure 1.4). The desire for both collaboration and efficiency is a contradiction that can be eliminated with the simple addition of 5%-10% built into the grossing factor.

Since programmers generally establish budgets for appropriations, bond issues or fundraising, it is not unusual for these people to create a trap where the available money is cast in stone based upon overly frugal space planning. It is common for insufficient funds to be allocated to create a facility which will meet the user's needs. This situation may be further exasperated by construction inflation between the time of budget preparation and the midpoint of actual construction. In the best of times for owners, materials and labor rise at some relatively small rate, but during construction booms inflation can be devastating. For example, in 2013, a moderately slow market, the magazine *Engineering News Record* shows costs increasing at just under 2.5% a year. If the budget setters have not recognized that at a minimum, it will be 4 years between initial discussions and the actual midpoint of construction then, somehow, somewhere, another 10% must be taken out of a project resulting in less space or less quality. Not long ago, in 2007, construction costs in Las Vegas were escalating and compounding at 1.5% per month. This was nearly 20% on an annualized basis. It's easy to see in that case, over a 4 year period, planners needed to double project budgets just to account for rapidly rising costs and contingencies (Figure 1.5).

"由建筑师来确认设计任务这点非常关键。这不仅仅是因为建筑师可以就建筑的需求和项目的利益与相关者进行重要的探讨,也因为设计任务的制定必须要考虑到实际的预算条件。"

"Program confirmation by the architect is critical. Not only does the architect begin important discussions regarding building needs with the stakeholders but this is when the realities of budget sufficiency must be addressed."

关于亚利桑那大学诗歌中心设计任务的确认过程包括了关键决策人、设施员工以及诗人代表,大家一起讨论决定项目的规模和预算。

Program confirmation for the University of Arizona Poetry Center included key decision makers as well as facility staff and poets to determine the project scope and budget.

通过上面的讨论可以清楚地看出当一个建筑师被选择去按照一个业主提供的设计任务来设计的时候，建筑师能否积极地参与到设计任务的确认过程是非常关键的因素。

在确认设计任务的过程中，建筑师不仅仅需要和项目的使用者讨论建筑功能的设定是否合理，而且预算是否实际也必须被加以讨论。能够反映出通货膨胀因素，突发事件以及实际空间需要的合适的补偿系数需要被加入到计算当中，如果由此得出的数值同最初计划的预算存在较大的差异，那么计划就必须进行调整，产生的财政方面的问题必须同项目决策人一起去解决。通常来说，解决方案有两个：一是设法获得更多的资金；二是缩减项目规模。在多渠道融资项目的情况下，例如发行债券或是联邦政府计划的项目，一旦出现上述情况，项目很有可能会半途而废。

有一个方法可以减轻预算不足带来的危害，特别是有压力需要使用廉价的材料和低效的系统来节约初始投资成本时，那就是寻找出原先整体的项目计划的一部分可以晚一些（即推迟到下一个阶段）去建造。这样的方法使得业主可以花更多的钱在更耐久的部件和更高效的系统上。当需要建筑的维护和运营成本更经济的时候，这是个不错的方案，因为筹募建造建筑的资金（资本募集、发行债券等）总是要比筹募后期运行建筑的资金（运行、维护建筑的预算）要容易得多。

非常有趣的一点是，当建筑被个人出资捐赠建造时，如果可以针对经费的短缺问题对出资人做出合理的解释，他们通常都会追加经费，而不是通过建筑设计去进行妥协。

From the discussion above, clearly it is critical that, when an architect is selected to design an owner programmed facility, the architect engage in a **Program Confirmation Process**.

During this validation procedure, not only does the architect begin important discussions regarding building needs with the stakeholders, but this is when the realities of budget sufficiency must be addressed. Appropriate grossing factors showing inflation, contingencies and real space requirements, if different from preliminary planning, will be disruptive, and the resultant financial problem must be resolved with key decision makers. Normally, the choices available are obtaining additional funding or reducing project scope. In the case of multi project funding such as a bond issue or federal programs, whole projects may fall by the wayside.

One methodology that can help mitigate budget shortfalls, particularly when there is pressure to use cheap, low performing materials and less efficient systems to save first costs, is to identify certain aspects of the program which can be made an integral part of the project design, and plan that they be constructed or finished in a later phase. This approach allows spending more money on long lasting components and high efficiency systems. It is building for economy of maintenance and operations as a first concern that is congruent with the idea that it is much easier to raise money for constructing buildings (capital campaigns, bond issues, etc) than it is to fund their operations (operations/maintenance budgets).

It is interesting to note that in the case where projects are gift funded, that is individuals have donated money to construct a building, it is not unusual, when presented with well reasoned explanations of a budget shortfall, that donors will give additional money rather than see the project compromised.

读者可以看到不经过建筑师审核以及使用者参与讨论并达成一致就盲目地去接受一份业主提供的任务清单可能导致的一种最坏的结果。预算的不足以及空间分配的不合理会导致十分严重的问题。同样糟糕的事情是不能就使用者的目标及需求得到他们第一手的确认。在很多案例中,一些相互冲突的设计目标,例如在之前章节讨论的那个有着倾斜的玻璃幕墙的建筑,在建筑师同使用者讨论之前是不会被写进任务文案中,甚至是不被知晓的。

可以进一步证明建筑师参与论证设计任务重要性的是这样一个案例,我们要设计一个新的诗歌中心建筑,在确认设计任务的过程中,建筑师就客户的需求提出了一系列矛盾的设计目标来同客户进行交流。这些相互矛盾的设计目标只有通过客户和建筑师之间大量的对话才能被揭示出来。对于一些关键冲突点的调和不仅仅引导出了有意义的建筑设计,而且其中的一组矛盾点(图1.6),即看书时对于自然光的渴望和避免书籍暴露在直射的阳光下从而受到紫外线的伤害之间的矛盾,启发我们发展出了一套新型的书籍管理系统(图1.7)。

作为所有设计工作的基础,建筑师在开始设计之前都必须对将来项目所在的场地进行深入的了解。在确认设计任务的过程中,设计团队将会完成一份详细的场地分析从而可以对于项目场地的环境、地质以及文脉等信息有一个全面的了解。

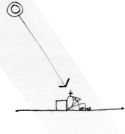

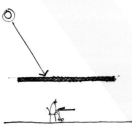

(图1.6)设计任务讨论过程中绘制的设计矛盾点示意图展示了阅读书籍时对于自然光照的需要同保护书籍免受自然光中紫外线伤害之间的矛盾。
(Figure 1.6) Contradiction diagram developed during programming illustrating the ability to read in natural light VS protection from UV rays.

The reader can see that blindly accepting a client furnished program, with no agreement as to architect review and stakeholder input, can lead to the worst kind of situation. Insufficient budget and space allocation will cause severe problems. Equally bad is not having access to stakeholders to confirm, first hand, their goals, needs and wants. In many cases conflicting goals, such as those mentioned in the angled glass building discussed in prior paragraphs, will not be documented or even known unless the architect meets with the users to verify the program.

Further supporting the importance of the architect verifying a planning groups program is the situation where a number of major contradictory goals were communicated regarding requirements for a new Poetry Center Building during this confirmation process. These inconsistencies could only have been discovered through extensive dialog between the clients and the architects. Not only did reconciling what were essentially competing interests lead to meaningful architecture, but one of the contradictions (Figure 1.6), the desire to have patrons read poetry books in natural light while limiting their exposure to harmful ultraviolet rays, inspired the development of a new type of book management system (Figure 1.7).

Fundamental to any design, the architect must gain an in depth understanding of the place the project will be. During the program confirmation process, the design team will complete a detailed site analysis and gain a full understanding of the environment, geographic and contextual issues.

设计任务的制定
Programming 27

书籍管理系统示意图
Book Management System Diagram

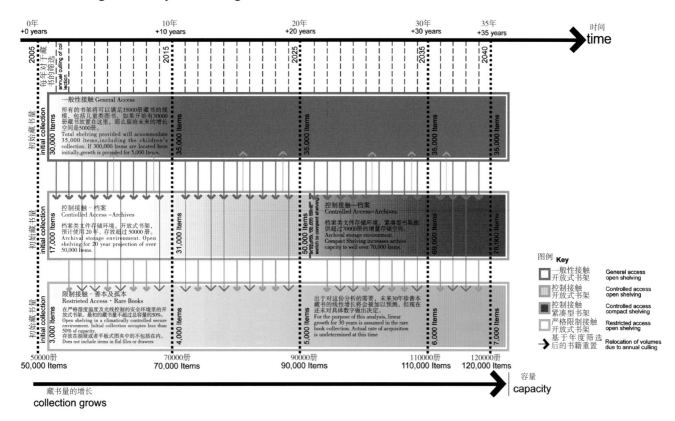

图1.7 在设计任务制定环节发现的矛盾导致我们发展出一套书籍管理系统

以在设计任务讨论会上收集到的信息为基础,建筑师发展出了一套诗歌中心书籍管理系统——这解决了保护书籍同允许在自然光照下阅读书籍之间的矛盾。这个想法类似于通过将艺术品保存在一个恒温恒湿的环境中来保护死海古卷。在诗歌中心,在书籍出现损伤的征兆前,都会被允许在自然光环境下阅读。一旦发现有破损的迹象,它们就会像死海古卷一样被移至一处人工控制的环境中加以永久保存。

Figure 1.7 Contradictions during programming led to the development of a Book Management System

The Poetry Center Book Management System was proposed by the architect based on information gathered during programming workshops - this solved the contradiction of preservation of books while at the same time enabling them to be read in natural light. The idea was analogous to the preservation of the Dead Sea Scrolls by keeping those valuable artifacts in a perfect, climate controlled environment. At the Poetry Center the books would be seen in natural light until they began to show signs of deterioration, at which point they, like the Dead Sea Scrolls would be moved into a climate controlled environment so they could last indefinitely.

在设计任务讨论中发现的互相矛盾的设计目标包括同时需要独处和互动。通过空间的收紧、室外花园的设计很好地解决了这一矛盾，一侧宽阔的空间允许社交互动，另一侧收紧的空间则适合独处。

Contradictory goals during programming included the simultaneous need for both solitude and interaction. The garden space addresses this contradiction by tapering to allow social interaction on the large side, and solitude on the other.

建筑师主导制定出设计任务

项目前期同使用者的讨论会帮助我们建立功能需求，同时草拟出设计目标和规则，而这种定义项目设计任务的方法（即建筑师主导制定出设计任务）将建筑师置于这个对话的中心。最先开始的关于预算的讨论经常会关注一些实际的财政问题，例如筹募的资本和建筑运营的费用（包含建筑维护的成本）。基于各种不确定性，一些不同的成本构成都会作为设计任务讨论的一部分，包括不同的建设分期以及项目规模的调整。重要的是，场地、文脉和环境等因素会成为这部分讨论的核心内容。当讨论中出现有争议的话题时，参与者可以通过协商来达成共识。在这个阶段就可以提前发现潜在的问题是很有价值的，远好过在完成了一个有缺陷的设计后才意识到有些事情行不通。有意思的是，设计任务中的一些互相矛盾的要求经常可以引导出一些富有创造性的解决方案。

或许能够直接参与设计任务的制定这种方法最大的好处就在于设计师对于建筑功能及环境的需要是如此的熟悉，于是当设计向前推进时，由于他们是任务拼图的制定者之一，设计概念就会很自然地被揭示出来。

如果设计师从一开始就参与到项目任务的讨论，一些超出预期的、有趣但却又是非常关键的信息往往会被揭示出来，例如对于关键问题的识别。一些看起来似乎是很小的事情，如开关插座

Architect Directed Programs

This approach to project definition places the architect in the middle of the dialog establishing needs and wants and concurrently allows a joint crafting of goals and precepts with stakeholders. Budget discussions are at the forefront and often address the realities of capital funds vs operational expenses (including maintenance). Given uncertainties, various cost scenarios will be presented as part of the program including phasing and scope modification. Importantly, place, including site, context, and the environment becomes a central part of these discussions. When competing needs arise during discourse, consensus among stakeholders can be negotiated. At this early stage it is valuable to identify potential problems rather than after completion of an ill-informed design. Paradoxically, conflicts within the program can lead to creative design solutions.

Perhaps the biggest benefit of participation in programming is that the designer becomes so connected to the issues and so informed about the building and environmental requirements that as the process begins to wind down, the building concept starts to reveal itself as the natural and obvious solution to the puzzle which he or she has helped create.

Interesting but crucial types of information, beyond the expected, tend to be revealed when the designer is involved from the beginning. One example is the identification of critical issues. Seemingly small things such

的位置或是某个特定设备所需要的空间，在使用者的眼中都有可能变成项目成败的关键。另外一个经常会被突然发现的事情是出现个人或是团体形式的项目的反对者。设计师有能力直接去讨论他们所关注的问题，而不是通过业主的过滤或是带有某种偏见的阐述才能获取相应的信息，这样他们就可以真实地去评估这些问题是否可以得到解决。

同制定设计任务一样重要的是获取设计需要的基础性的知识，同等重要的是，通过直接参与这个过程，建筑师可以开始识别出一些未知或是待解决的问题。通常这些因素尽管对于最终的结果甚为关键，但并不妨碍最初工作的展开，而仅仅是设计任务中诸如"在X可以继续之前我们需要先知道Y"这样的表述。我们可以用线和空间事务所设计的拉斯维加斯旁边的沙漠学习中心项目来做个例子，在设计任务的讨论过程中，一些会对设计产生影响的不同要素被列了出来，例如制定一个能够反映出功能要求的正规的学区协议，建立一个管理委员会，以及解决水源问题，这样我们就可以在多种可能性之间做出一个选择。

在亚利桑那索诺兰沙漠博物馆餐厅及画廊综合体项目中，当设计任务被编订出来后，我们可以很明显地看出来，如果仅仅是去维护旧有的维圣达菲建筑风格，而不是让功能美学需要以及自然环境的力量去塑造出建筑形象，项目将面临很大的压力。当考虑到沙漠博物馆经常以自己是动物园设计的先锋为荣的时候，这一点就尤为重要。为了保持同自己形象的一致性，沙漠博物馆的

---

as location of electrical receptacles or space for certain equipment may in the users' eyes be the difference between a buildings success and failure. Another important aspect, often abruptly revealed, is the identification of naysayers, that is, persons or groups that are against the project. The designer has the ability to discuss their concerns directly, instead of receiving filtered or prejudiced re-statements from the client, in order to see if these matters can be resolved.

As important as programming is for gaining the knowledge fundamental to the design, equally important, by participating directly in this process, the architect can begin to identify unknowns or issues to be resolved. Often these items although crucial to final outcomes may not obstruct initial work but are simply part of the statement that before "X" can continue "Y" must be known. Examples of this occurred during programming for the Desert Learning Center near Las Vegas and included such varied items as formalizing school district agreements reflecting programming decisions, establishing a governing board, and completing water source feasibility so that a choice could be made among various options.

In the case of the Restaurant and Gallery Complex at the Arizona Sonora Desert Museum (ASDM), as the program was developed, it became apparent that pressure was developing to maintain the appearance of an established pseudo Santa Fe Style of architecture, rather than let the form and appearance be shaped by the functional and aesthetic needs of the museum tempered by environmental forces. This was a particularly important issue because the ASDM prided itself on being progressive and at the cutting edge of zoo design. To be consistent, Desert Museum Architecture should be of its time and place, modern and respect-

建筑就应该顺应所处时代和场地的要求，现代并且尊重当地的文化和环境。设计任务文件就是一个非常理想的将这一问题表达出来的地方。下面标题为《意象》的文章摘录自这个项目真实的设计任务文本。

意象

沙漠博物馆餐厅的建筑应该传递出一种什么样的视觉上的美感？这样一个会引起很多争论的问题，其实有一个简单的答案，我们有机会去创造出一个可以对环境做出恰当回应的沙漠建筑范例，而不是试图通过重建一些虚假的舞台布景一样的建筑或是类似于外凸的梁架这样的传统符号来显示我们的敏感性。

一个建筑必须适合它所在的场地。它必须要属于那个地方。这个餐厅建筑应该"安静"地追随先前建造的沙漠博物馆建筑物，但这并不意味着我们的设计要去抄袭之前的那些建筑，重要的是我们应该保持那些建筑所展现出来的对人体尺度的敏感性。沙漠博物馆里的建筑都很低调，周边生物产生的阴影在它们所形成的屏幕上起舞。它们不是为吸引人的眼球而生的。它们从不大声喧哗。它们都同周边环境有着"柔软"的接触。

---

ful of the culture and the environment. The program was the ideal place to address this issue head-on, as you can see from the following excerpt, titled *Imagery*, from the actual ASDM Program.

Imagery

What visual aesthetic should the Desert Museum's restaurant buildings convey? A question that could provoke much debate has a simple answer; we have an opportunity to create an example of appropriate desert architecture that relies on environmental response as its primary determinant instead of fanciful stage sets which purport to fit the desert by recreating miniature missions or constructing foamed beam protrusions as the answer to sensitivity.

A building must fit the land. It must belong. The restaurant architecture should be "quiet" following the precedent set by other ASDM structures. While there is no need to make a copy of these buildings it is important to follow their lead in sensitivity to human scale. ASDM buildings are uniformly low. They form screens for the projection of biological shadow plays. They do not compete for attention. They do not shout. They are "soft" edged.

  光与空间作为设计的要素必须成为建筑形象重要的组成部分。这一策略在博物馆的入口处通过巧妙的设计被加以放大，格栅型滤光栅将刺目的阳光柔化为戏剧性的光影的舞蹈，同时框出一个无与伦比的深远的景观。入口好像在说："欢迎来进行一次探险。"我们当然可以看到当地文化上的影响，古印第安人居住区精妙的几何造型、帕伦克的力量以及早期居民在建造时对于"A"字山上玄武岩的运用，所有这些都影响到了我们的思考。但是，这里是亚利桑那索诺拉沙漠博物馆，不是新墨西哥州奇瓦瓦沙漠博物馆或是玛雅神庙。

  在许多人的脑海中墨西哥的殖民主义建筑形式可能会产生出最理想的浪漫画面。但我们必须记住我们不是在依赖神话故事去售卖房产。我们应该去真正理解墨西哥建筑的精髓，而这其实同钟塔没有多大的关系，它的一切都是为了表达愉悦与大胆、毫不妥协的颜色（松石绿色的门道、艳红色的侧墙——而黑色根本就不能被考虑），鲜活的就好似图森市生动的日落。这是一种以墙作为主导性的关键结构元素的建筑。这是一种由充满着美景、神奇与惊喜的入口序列感引领你进入圣境的建筑，一条被预先安排的道路会诱惑你去揭示最后的高潮。

Light and space are design elements which must be part and parcel of the architectural image. This is skillfully exemplified at the entry to the museum where a lattice filter tempers the harsh sun with a dynamic dance of shadow while simultaneously framing an incomparable distant view. The entry gently says "welcome to an adventure". Clearly there are cultural influences to be recognized: the masterful geometry of Pueblo Bonito; the power of Palenque and the early settlers use of "A" Mountain Basalt all temper our thought, however, this is the Arizona Sonora Desert Museum not the New Mexico Chihuahuan Desert Museum or a Mayan Temple.

The romantic idealization of Colonial Mexico probably forms the strongest associative imagery in many minds but it must be remembered that we are not in the business of selling tract homes based upon myth. We should be proponents of a true understanding of the essence of a Mexican architecture which has little to do with bell towers and everything to do with the joy of bold, uncompromising color (turquoise doorways, bright red nunnery walls - dare you think black!) as shameless as the most vivid of Tucson sunsets. It is an architecture where the wall is key as the dominant structural component. It is an architecture of beauty, mystery and surprise that develops the entry sequence to rival the best; you are coaxed along a determined path of temptation to a revealing final climax.

自然的材料在设计中必须扮演重要的角色，它将结构同土地有机地联系起来。建筑如何同天空接触将会决定视觉上的成功与否。这一点在沙漠博物馆项目中尤其重要，因为我们的任务不仅仅是教育参观者了解周边生物的复杂性与多样性，也需要让他们知道做一个好的沙漠居民的意义所在，建筑可以帮助传达这个信息。

就像一些陈词滥调说横穿建筑侧壁的梁总是会同结构直接冲突一样，这样的事实提醒我们人们总是会对他们既有的现状不满，总是希望成为一些什么别的东西。希腊人试图让石头看起来像木头；而文艺复兴时期的人，理想化地使用白色的抹灰来让他们的建筑看起来像他们无法负担的石材，却完全忽视希腊人早已被忘记的本意和原本他们神庙上鲜丽的色彩。我们所谓的创造性思维经常会产生一种反常的连续性，试图让塑料看起来像木头，或是让木头看起来像塑料。在此同时，一个时代的感叹："我们所追求的诚实到底怎么了？"仅仅是一个反问句吗？难道我们的项目不应该成为答案的一部分吗？

建筑形象必须是来自于我们的灵魂深处，而不是既定的处方。我们通过理解合适的材料、气候、场地和需求去寻找答案。

Natural materials must play an important role in the organic relationship that welds structure to earth. How the building touches the sky will determine visual success. This is particularly true at the ASDM where our mission is to teach not only the wonders of biological complexity and diversity; but the meaning of good desert citizenship. Architecture can help convey this message.

Clichéd beams poking through the sides of buildings often in direct contradiction to structure remind us that humans have always been malcontent with what they are and always hope to be something else. Greeks tried to make stone look like wood; Renaissance man, ignorant of the garish polychrome of the temple and the Greek's long forgotten goal, idealized, in white plaster, what was supposed to look like the stone they could not afford. Providing a perverse thread of continuity our creative minds try to make plastic look like wood and wood look like plastic. Meanwhile, a generation laments "what ever happened to honesty"? Is it just a rhetorical question? Isn't our mission to be part of the answer?

Architectural image must come from the soul not from prescription. We help it along by understanding the appropriate materials, the climate, the land and the need.

设计任务书是向使用者介绍建筑师设计哲学的一个有力工具。当客户真正参与到设计任务的制定过程中之后，他们就可以很容易地理解肤浅的风格标签与建筑应该属于它的时代和场地这样有意义的想法之间的区别。实际上，一旦设计任务书被拥有否决权的决策者确认通过，就意味着在设计任务书中被清晰地表述出来的建筑师的设计哲学已经被甲方很好地接受了。

交织在这一章节中的说明性的例子展示了在引导出最终的设计成果之前（图1.8、图1.9）如何通过设计任务书来描述建筑并揭示出一个粗略的概念示意图（图1.10）。

让设计师全程参与设计任务书的制定过程的最后一点好处是，在获知了实际的预算以及经常出现的同实际需要相比过多的要求时，建筑师可以参与将这些目标排列出一个优先级别。这就意味着建筑师可以将诸如场地及建筑设计对它的回应、节约能源、材料的选择、对自然光的运用、空间的质量、设计应对未来变化的适应度及灵活性、长期的运行成本等一系列的问题按照优先级别排序处理。

在研究过这章关于设计任务制定的介绍之后，我可以确定本书将会专注于讨论建筑师参与制定设计任务。

The program is a powerful tool for introducing stakeholders to the architect's philosophy. When the client has been immersed in the programming process they easily comprehend the difference between superficial labels of style and the more meaningful idea that architecture should be of its time and place. The fact that the Decision Maker has, in fact, veto power over the information documented and explained in the program means that once the program is approved, the architect's design philosophy, which has been clearly expressed, is well accepted.

Interlaced within this section are illustrative examples which demonstrate how the program was used to shape the building informing a rough concept diagram (Figure 1.8, 1.9) before guiding the design to its final outcome (Figure 1.10).

One last point in favor of the immersion of the designer in creating the program is that, given the realities of budget and the often overpowering nature of wants compared to needs, the architect is able to participate in establishing priorities. This means that the site and the way the building responds to its forces, resource conservation, material choices, integration of daylight, quality of space, flexibility for future adaptations and long term operating costs all have the architect to advocate for their appropriate place within the order of things.

After studying this introduction to programming, I am sure that it has become evident that informed will focus on Architect Directed Programming.

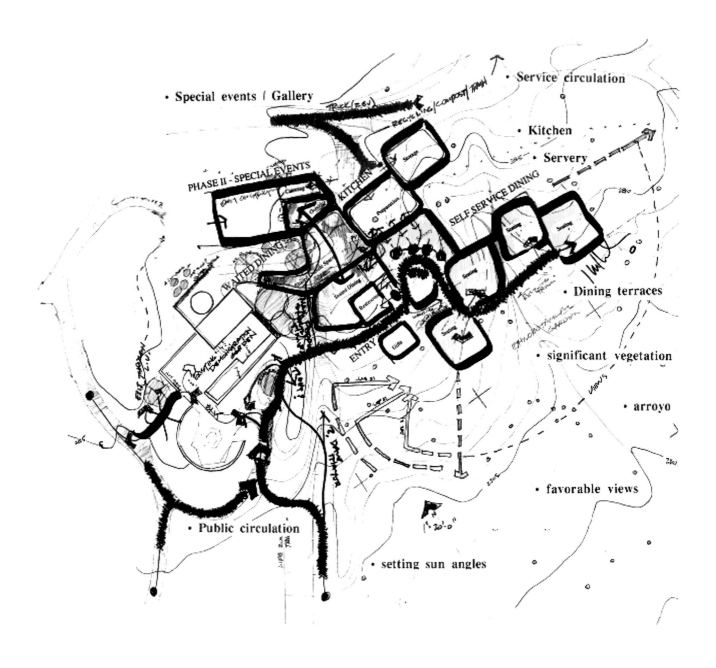

（图1.8）这个最初的示意图展示出了如何在设计任务制定的过程中去回应一些关键的问题，例如地形、对原生植被的保护、功能空间邻近关系、客人同服务流线的分离、景观视线、建造分期、现有结构以及当地气候等，它们共同塑造出最终的设计方案。

(Figure 1.8) This initial diagram begins to reveal how responding to critical issues identified during programming such as terrain, preservation of native vegetation, functional adjacencies, separation of visitor and service circulation, views, building phasing, existing structures, and climate begin to shape the design solution.

(图1.9)这张平面示意图专注于分析景观视线,在最初的示意图的基础上有所发展,但仍然关注到了设计任务中一些关键的问题。

(Figure 1.9) This plan diagram, emphasizing views, shows development from the prior diagram while still addressing critical program issues.

(图1.10)亚利桑那沙漠博物馆项目的建筑形式来源于功能要求以及它所处的环境,而不是先入为主的风格符号。
(Figure 1.10) Instead of a building conforming to a preconceived pseudo style, the ASDM takes its form from the program and its surroundings.

设计的开始
Beginning

这一章节阐述了如何开始制定设计任务的过程。在这里,我们在研究场地或是计划一场同项目使用者的对话之前,会先去检验我们要做的事情。

This section explains how to start the programming process. Here we examine things to do before studying the site or structuring dialog with stakeholders.

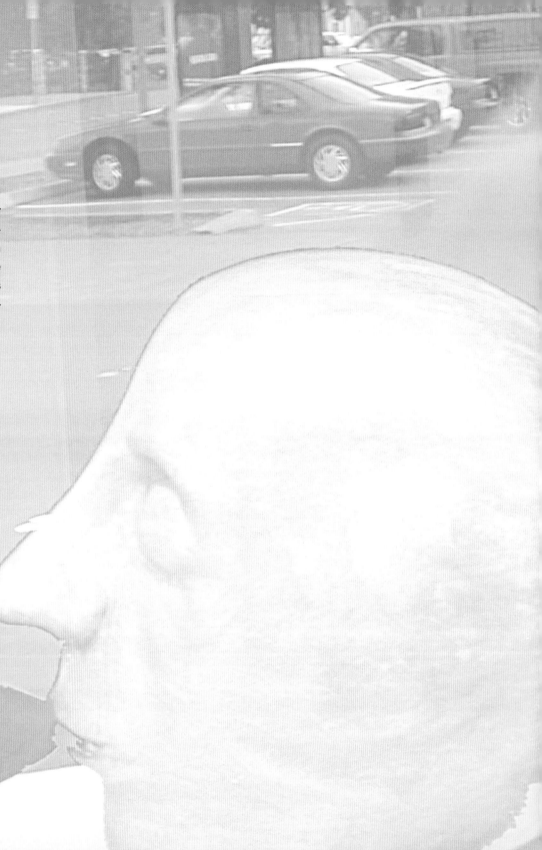

（图2.1）背景照片：诗意的（想象）或是务实的（停车的需求），项目决策人的愿景决定了设计的走向。我们为什么要建造这个项目？哪些是关键性的问题？怎么来定义项目成功？
(Figure 2.1) Background photo...Poetic (imagination) or pragmatic (parking requirements), the DM's vision for the project informs the design. Why are we creating this project? What are the critical issues? What defines success?

设计的开始
Beginning

　　在建筑师被选中设计一个新的项目之后，早在进行商务谈判的过程中，建筑师就应该同项目的决策人安排碰面开会；所谓项目的决策人就是指被授权可以确认项目设计的人。

　　在这个碰面会上，应该制定出一个带有下列目的的议程表（参看65页典型的议程表）：①明确地理解客户布置的任务；②确定谁是项目利益相关者（关键参与人）以及他们是否有时间参与；③安排重要会议的日期和时间；④决定会议的框架和设置（形式和场地）；⑤定义用户并协助获得长期用户的信息（对于访客的分析等）；⑥收集现有的文件（从组织结构图、工作职责描述到环境研究报告等所有与项目有关的可用文件）；⑦获得长期的场地信息，包括水文及土壤分析报告、考古研究、环境研究（环境评估以及环境指标）和一份完整的工程测量报告，包括场地边界、地形、地役权、公共设施、场地出入方式和原有结构物；⑧讨论希望的项目总工期；⑨项目预算；⑩取得项目决策人对于项目看法的第一手资料（图2.1）。

After the architect has been selected to design a new project and as early as practical in the contract negotiation process, the architect should meet with the Decision Maker (DM); the individual who has the ability (possibly, subject to Confirming Authority authorization) to approve the project design.

For this meeting, an agenda (See Typical Agenda pg.65) should be developed with the desire to: ①Gain a clear understanding of the client's mission, ②Identify stakeholders (key participants) and their availability, ③Schedule dates and times for primary meetings, ④Determine the meeting framework and setting (format and place), ⑤Define users and facilitate the procuring of long-lead user information (visitor analysis, etc.), ⑥Gain access to readily available documents (everything from organization charts and job descriptions to title reports and environmental studies), ⑦Expedite long lead site information including hydrology and soils reports, archeological studies, environmental studies (EA and EI) as well as a complete engineering survey including boundaries, topography, easements, utilities, access/egress and existing structures ⑧Discuss the hoped for total project duration, ⑨Budget, and ⑩Gain first-hand knowledge of the Decision Maker's vision for the project (Figure 2.1).

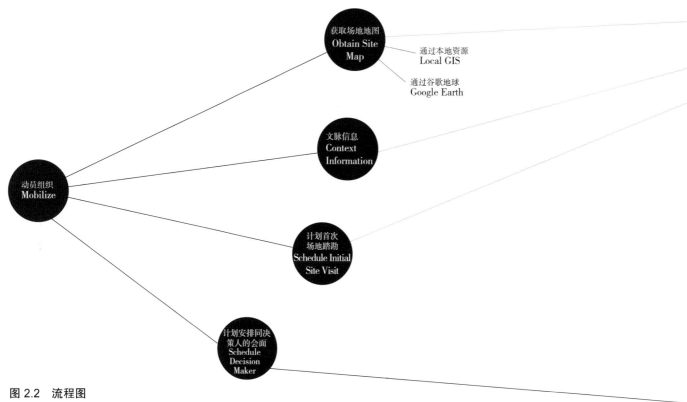

图 2.2　流程图
Figure 2.2 Flow Chart Schedule

早在收到项目合同之前，建筑师就应该绘制一张流程图，这其中应该包括一系列在设计师主导的对话开始之前就必须完成的活动。从最开始的动员、与项目决策人安排会议日程，直到收集各种相关的场地和项目利益相关者的信息。流程图是一份非常有价值的示意图，它会标示出长期项目，并且帮助建筑师在设计任务制定初期发现一些潜在的冲突点。本页的流程图描绘了几个设计任务制定过程开始阶段发生的活动之间的关系。将所需时间标注到不同的任务阶段会让流程图变得更有价值（图2.2）。

As early as feasible after receiving a contract, the architect should begin to layout a flow chart schedule that includes a network of activities that must be accomplished before designer directed dialog can begin. From initial mobilization and scheduling a meeting with the DM, to gathering relevant site and stakeholder information, the flow chart will be a valuable diagram to identify long lead items and to help the architect identify potential clashes early in the programming process. The flow chart diagram on this page depicts several typical relationships between activities that take place at the start of the programming process. Time can be applied to each of the phases to make the diagram even more valuable (Figure 2.2).

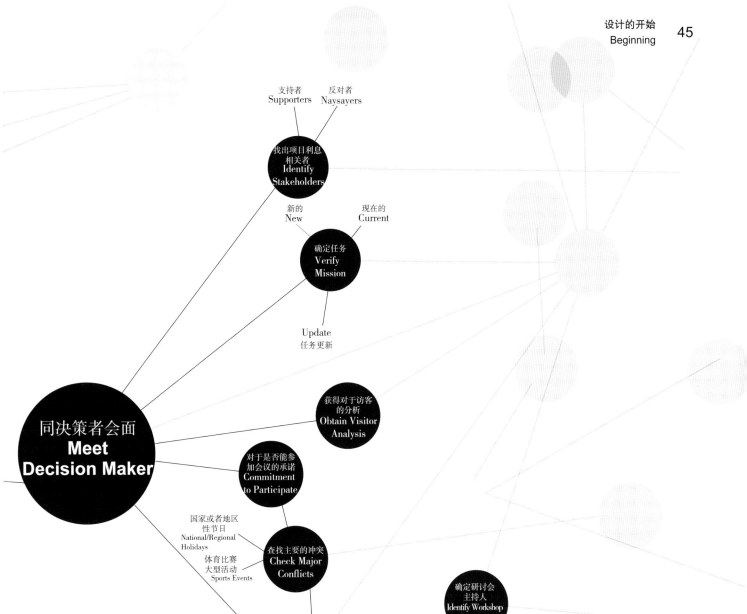

任务宣言

一个任务宣言阐释了组织机构存在的原因。依照大卫·库兰为美国建筑师协会所写的书中所言："宣言指引着行动的方向。"宣言由精确的语言以正规的格式构成，建筑师通常可以比较容易获得这样的资料，例如通过发表在时讯上的文章、组织条例、网上的信息或是通过其他的渠道。因为组织机构制定发布任务宣言是经常发生的事情，项目的决策人应该向建筑师明确所提供的任务宣言是否是最新的。在项目的中心目标还不明确、任务宣言尚未完成的情况下，有关设计任务的对话应该被推迟。

在美国，可以通过法律来设置一份任务宣言，例如美国森林服务局"在可持续多用途管理的理念下实现高质量的土地管理以满足人们的不同需要"；也可以通过公司的董事会制定出来，例如耐克公司的"将灵感与创新带给世界上的每一个运动员"；抑或是通过一个非营利组织编写出来，例如美国红十字协会的"在紧急事件发生时，美国红十字协会将会通过组织志愿者及慷慨捐赠的方式来预防和减轻人类的痛苦"。

任务宣言将会是一个指引，就如同建筑展开的定义一样；它不仅指出了机构所追求和代表的利益，而且，同等重要的是，它还表明了机构所反对的事情以及机构所计划的新建筑应该或不应该追寻的方向（图2.3）。另外，一系列附带由甲方机构编写的被称之为"愿景声明"的指导意见，也会对建筑师理解客户的核心价值有所帮助。

Mission

A mission statement articulates an organizations reason for existing. According to David Koren in his book for the American Institute of Architects, *"the mission guides the direction of the action"*. Formulated in precise language, this declaration is normally readily available to the architect; published in newsletters, organization by-laws, online and a variety of other easily accessible places. Since it is not unusual for a group's mission to evolve, the DM should verify to the architect that it is current. If this overarching goal has not been crafted or is not clear, the dialog aspect of programming must be delayed until the mission statement is completed.

In the United States, a mission statement may be set by jurisdictional law as is the case with the US Forest Service: "*to achieve quality land management under the sustainable multiple-use management concept to meet the diverse needs of people*"; created by the board of a corporation such Nike to: "*bring inspiration and innovation to every athlete in the world*"; or a nonprofit organization such as the American Red Cross: "*The American Red Cross prevents and alleviates human suffering in the face of emergencies by mobilizing the power of volunteers and the generosity of donors.*"

The mission statement will be a guide as the building definition unfolds; not only indicating what the organization stands for but also, equally important, what it is not and by implication what the group's new building is and is not (Figure 2.3). An additional series of guiding precepts, developed by the client organization and known as the "vision statement", may also be available to help the architect understand an organization's core values.

（图2.3）亚利桑那大学诗歌中心集中了几处多功能的室内室外聚集场所，可以容纳多种活动。这个概念来自于诗歌中心的任务描述："弘扬诗歌文化并且保持，丰富和推进多元化的文学土壤。"

(Figure 2.3) The University of Arizona Poetry Center integrates several multipurpose indoor and outdoor gathering areas that can be host a variety of events. This concept was guided by the Center's mission statement, *"to promote poetic literacy and sustain, enrich, and advance a diverse literary culture."*

利益相关者

项目的利益相关者，是指那些与项目有利益关系的人，范围包括从项目所有人、公司董事会成员、项目使用者、访客、雇员、志愿者、支持团体、项目捐赠人、个人、邻居直至居民委员会。这份清单还可以延续，包括公共设施代表、维修部门、法规制定部门及政府代表，和一些特殊群体（例如企业支持者，环境保护组织，例如希拉俱乐部，工会组织例如教师联盟等）都可能被包含在内。很显然，根据项目的不同，利益相关者的类型和数量都会有很大的不同。

最开始的问题之一，就是询问项目决策人谁是关键的项目利益相关人，并使之能够参与到会议过程中。重要的是不只是姓名，客户还应该提供利益相关人的其他联系方式，包括电话号码、电子邮箱、邮寄地址等。这个团体会随着时间的流逝而不断增加，但在项目开始时迅速辨识出关键利益相关人可以让他们尽早地参与到讨论的过程里。

组织架构表会清晰地显示出一个机构内部的组织层级以及管理方式。建筑师应该同项目决策人一起审核这份文件以确认未来项目讨论会的参与者（和可能的主持人），建筑师还可能希望得到此研究机构中所有职位的岗位职责。这样的工作不仅可以让建筑师对组织机构中人员的责任和相互之间的关系有一个更好的了解，也为将来收集个人信息时所需要准备的有意义的问题构建了坚实的基础。

Stakeholders

The stakeholders: persons who have an interest in the project, range from owners and boards of directors, to users and visitors, employees and volunteers, support groups and donors, individuals, neighbors and homeowner associations. The list goes on to includes utility representatives, maintenance departments, regulators and government representatives and special interest groups (i.e. Business proponents, environmental protection advocates such as Sierra Club, Labor representatives like Teachers' Unions, etc.). Obviously, depending upon the project, the type and number of stakeholders vary dramatically.

One of the first questions, is to ask the DM to identify key stakeholders so they can be invited to participate in this process. It is important that in addition to names, contact information including phone numbers, e-mail and street addresses are provided by the client. This group will evolve, growing over time, but prompt identification will allow early notification in order to facilitate participation.

A clear picture of an institution's internal hierarchy and the way it is administered will be shown by their Organization Chart. The architect should review this document with the DM to identify participants (and possible moderators) for future workshop sessions. Additionally, the architect may want to obtain and study job descriptions for all of the positions within the group. This not only gives the architect a better understanding of relationships and responsibilities among people within the organization but forms a basis for formulating meaningful questions for use when soliciting information from these individuals.

毫无疑问，可以立即得到项目决策人的承诺，说他（她）会全程参与设计任务讨论会并要求他（她）的下属也这样做，这具有非常重要的意义。当其他关键的项目利益相关人不能出席会议时，通常项目决策人可以说服他们参加。实际上，当项目利益相关者关注于项目的某些具体问题时，他们会有意愿来参加会议。让不同的利益相关群体都能参与到会议过程中是非常有必要的，这样可以保证项目的设计标准或目标达成一致，即便存在不同的意见，矛盾点也可以被清晰地展示出来。

积极参与并不意味着项目利益相关人需要出席每一次信息收集会，而只是需要持续关注讨论进程，在恰当的时机分享意见，并承诺预留出审核设计标准的时间。因为建筑师会直接根据设计任务来进行设计，这个过程有多么重要是不言而喻的。

在所有项目利益相关者中，最重要的是那些对新项目持有异议的人。这些被称为"反对派"的人可能有着各种合理的反对理由，例如害怕资源减少、担心交通拥堵、关注景观视线被遮挡，或者仅仅是因为他们是"别在我家后院建房子"派的成员，这些人不反对项目本身，只是不希望项目离自己太近。无论是哪种情况，能够尽早地找到这些人并邀请他们参与到项目讨论的进程中都是很重要的事情，他们也许可以就这些问题提出一些建设性的解决方案，或者只是指出问题，这也会对项目有所帮助。我们有一个真实的案例，一个社区反对在他们的旁边建造一个环境教育营地，因为他们担心当地的水源不足，不能够同时支撑他们所居住的小镇和一所新的学校，而

Not to be discounted is the significance of immediately getting a firm commitment from the DM to personally participate throughout the programming process as well as requiring her/his personnel to do the same. While other key stakeholders may not be able to be required to attend meetings, usually the DM can be persuasive in her/his request that they do so. The fact that stakeholders have an inherent concern with the project at hand means that they will desire to join in. This engagement of interested parties is essential in order to assure that there is unanimity in establishing project criteria or that, in the case of differences of opinion, conflicts are clearly expressed.

Participation does not mean that individual stakeholders need to attend every information gathering session, but rather that they continuously follow along in the process, share in appropriate meetings, and commit the time required for thoughtful review of design criteria. Since the design will follow directly from the program, it is self evident how important this is.

Among the most important stakeholders are those that are antagonistic to the new project. These persons, known as naysayers, may have valid reasons for being against the work such as fear of depletion of resources, creating increased traffic and blocking of views or they may simply be members of the "not in my backyard" faction; those who generally favor the project but simply do not want it near them. In any case, it is important to identify these persons as early as possible so they can be invited to take part in the process with an eye to offering constructive solutions to resolve their issues or perceived problems. An actual case where a complete community was against the creation of a nearby environmental education campus,

最终这个问题在项目任务讨论的过程中得到了解决。建筑师专门组织了一场针对他们所关注的问题的讨论会，解释说计划修建的学校将使用一种完全不同的水资源来满足用水要求，而不会去使用小镇居民赖以生存的地下水资源。在会议结束的时候，原本对立的局面转化成了一个积极的结果。在这个案例中有一点很有趣，项目的业主是一个政府机构，本着不碰"马蜂窝"的原则他们原本并不想去和这些反对者对话。

如上所述，跟我们的直觉不同，发现问题后找出项目的反对者，并积极地与之对话沟通是很有帮助的事情。另一方面，根据我们的经验，项目的反对者可能会比较固执或是不太理性。在第二个例子中，一群人反对一个计划沿着亚利桑那的圣佩德罗河修建的访客中心项目，因为需要延续数英里长的市政管线。起初这件事看起来是因为他们不想看到这些管线，但是我们发现承诺将管线埋入地下并不能安抚这群反对者。最后经过了几轮的讨论后我们才知道，原来本质上的问题是他们同市政公司之间存在着私人恩怨，而并不是反对我们的项目。理解这一点对我们来说非常重要，因为业主并不希望在同当地社区有冲突的情况下推进项目（图2.4）。

---

because of apprehension that there was insufficient water to serve both their town and the new school, was resolved during programming. It was explained, during a meeting specifically to address their concerns, that the proposed school would be utilizing an entirely different water source than the aquifer that the town relied upon. By the end of the session a confrontation was transformed into a positive experience. It is interesting to note that in this case, the client, a government agency that had been harshly criticized by the community, was reticent to meet with them under the theory that it was better not to disturb a hornet's nest.

As mentioned above, counter intuitively, it is useful to identify and interact with persons who may object to a project because of a perceived problem. On the other hand, as we have found out in the past, reasons to try and obstruct the design may be strictly personal or irrational. In this second example, a group was against a Visitor Center along the San Pedro River in Arizona due to the necessity of extending utility lines for several miles. At first it seemed that the objection was based on the unsightliness of the cables. An offer to place the utilities underground did little to placate these people. Finally, after several discussions, it became apparent that their issue was a personal matter with the utility company itself, not our project. This was important to understand because the client did not want the project to progress if it would antagonize the local community (Figure 2.4).

设计的开始
Beginning 51

"在所有项目利益相关者中，最重要的是那些对新项目反对的人。"

"Among the most important stakeholders are those that are antagonistic to the new project."

（图2.4）一个在圣佩德罗河岸中心项目场地附近的居民固执地坚持要使用光电能源，尽管这已经被获知是一种代价高昂的选择。通过更深入的讨论后我们发现这个人同当地的能源公司之间有争议，对项目的反对并不是客观表达个人意见，而是想借助项目报复能源公司。

(Figure 2.4) A resident near the San Pedro Riparian Center site was adamant about the use of photovoltaic power even though this was a cost prohibitive option. Further discussion revealed that this person had a disagreement with the local power company over a personal matter and was simply using the project to seek revenge.

排定主要的会议日程

在这个阶段项目决策人和建筑师一起来决定可能的研讨会以及其他一些用来确定项目目标和需求的会议的日程。当领导已经安排出时间以后，其他人员就会遵从。会议时间的分配是项目向前推进的第一步，并且是独立于项目的整体计划来完成的。

设定研讨会最明显的一个问题就是要确定相关利益人是否能够参加。在这种情况下，这些关键人的日程一般都排得很满，因此，和他们确定项目任务研讨会的时间就需要当机立断，事实情况往往是随着会议进程的推进，我们就会有更大的机会让更多的人参与进来（图2.5）。这种必须等到更多人方便参加的时候才启动项目任务讨论会的做法，同要借着选定建筑师的势头立即将项目向前推进的急迫性相冲突。这种想要保有这样的势头的渴望其实是同能尽早将项目完成这样的一个普遍想法相伴而生的。当然，所有正常的日程表都包含有一些限制条件，应该在起草的时候被关注，例如相互冲突的截止日期、休假、节日、社区和国家活动（会议、汇报日、体育比赛等）。

通常来说，对于中到大型的项目，应该为会议预留两周的时间，我们要意识到不同的个人可能在这段时间里只有某些时候可以参加讨论（图2.6）。

Schedule Primary Meetings

This is the time for the Decision Maker and the Architect to pull out their calendars and establish possible dates for holding the workshops and other meetings defining goals, needs, and wants. When the leader has committed time, others will follow. This allocation of time is a first step in moving the project forward and is done independently of creating the project master schedule.

One of the most obvious issues in setting the workshop sessions is availability of stakeholders. In this case, these key individual's time may be scarce and hard to obtain, hence, dates for programming sessions should be set immediately, influenced by the reality that the further in the future meetings are scheduled, the greater the chance that more participants will be available (Figure 2.5). This practical notion of delaying the start of programming until it is convenient to more people, clashes with the urgency to begin immediately in order to build upon the momentum created with the selection of the architect. This desire to preserve momentum is coupled with the near universal wish for completion of the project as quickly as possible. Of course, all of the normal calendar constraints including competing deadlines, vacations, holidays, community and national events (Conferences, reporting dates, sporting events, etc.) should be considered when drafting the schedule.

Generally speaking, for projects of medium to large scale, two weeks should be blocked out for meetings realizing that various individuals will only be available at only certain moments within that time frame (Figure 2.6).

设计的开始
Beginning

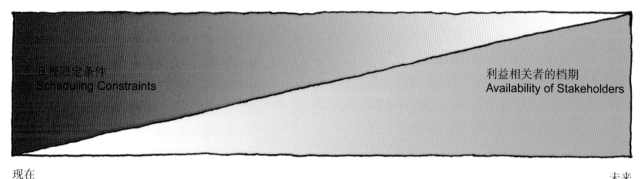

现在
Present

未来
Future

图2.5 与项目决策人开会,商定未来的日程
Figure 2.5 Meeting with the DM, Setting Future Dates

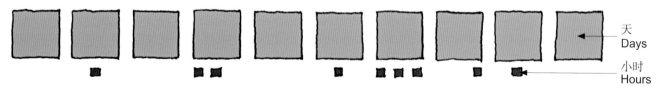

天
Days

小时
Hours

图2.6 计划时长 vs 设计任务制定实际需要的时间
Figure 2.6 Duration vs. Actual Time Required for Programming Meetings

会议框架或设置

在对话的过程中,当进入到了更深层次的关于会议框架或设置的讨论时,将一般组织会议的可能方式介绍给决策人就变得很重要。

为了同利益相关者建立对话而设定的会议在确定会议地点(或者几个地点)之前还有许多其他事情需要确定。它们包括预估将有多少人被邀请参会、这些讨论会的设置及谁会协助来做这些事。这些早期的决定对于最终的会议能否落地十分重要。

根据研讨所关注的主题和规模,讨论会应该由建筑师、外聘专家或者是甲方自己的专职员工来协助安排。

这种互动可以以正规的形式来组织(酒店的会议室和会议中心、礼堂、大学或是高中教室),也可以在非正式的场合出现(个人办公室、咖啡馆、图书馆里的社区活动室、社区学校,以及我们最喜欢的方式:在项目的实际场地搭个帐篷)(图2.7)。

可以组织一系列的讨论会,侧重的主体可以比较概括,也可以非常具体。会议的目的总是围绕一个新项目如何来融入场地、访客与员工的流线如何组织、建筑物内单独空间的需求如何界定来设定出目标或规则(经验上的、行为上的、环境上的、感觉上的、功能上的、操作运营上的以

Framework or Setting

While a more in depth discussion of meeting setting or framework occurs during the section on Dialog, it is important to introduce the general organizational possibilities to the DM.

There are a number of decisions to be made before establishing a place (or places) for creating dialog among stakeholders. They include establishing a rough estimate of how many individuals will be invited to meetings, the setting for these discussions and who will be facilitating. These early decisions are important to assure availability.

Workshops should be facilitated, depending upon the focus and size of the session, by the architect, outside experts or by client's staff professionals.

Interactions may occur in formal (hotel meeting rooms and conference centers, auditoriums and university or high school classrooms) or informal settings (individual's office space, coffee shops or cafés, library community rooms, neighborhood schools and our favorite; a tented space set up on the actual project site) (Figure 2.7).

The meetings and workshops are organized into a number of focus sessions with subjects ranging from general to very specific. The desire is always to establish **goals/precepts** (experiential, behavioral, environmental, sensory, functional, operational and budgetary), **needs** (capacity in terms of users, functions,

(图2.7) 一个在红石峡谷项目场地上支起的大帐篷被用来承办与项目利益相关者一起参加的设计任务研讨会。这种方式尤其适合于在关注场地问题的讨论会中创造出有意义的对话。
(Figure 2.7) A large tent on the Red Rock Canyon Visitor Center project site was used to host programming workshops with stakeholders. This method is particularly useful for generating meaningful dialog during workshops focusing on the site.

及预算方面的)、需要(在使用者、功能、家具及设备方面的要求)、希望(超出满足基本功能要求之上的需要)、邻近关系(功能如何定位以及相互之间的关系如何)和关键要素(如果少了这个元素,这个场所就无法取得成功)(图2.8)。研讨会期间,参会者会被要求书写或是绘制草图,所以一些类似于书桌或是会议桌的基本设施是十分必要的。

(图2.8)为了提升参与度,一个地区总监带领员工就面对一个新的设施所需要决定出的关键性问题展开讨论。
(Figure 2.8) In order to increase participation A Regional Director leads staff in an exercise that determines critical issues facing a new facility.

furnishings, equipment), **wants** (desires beyond satisfying basic function), **adjacencies** (how functions are positioned relative to each other) and **critical issues** (If ----- does not occur, the place will not be successful) regarding everything from how the new project meets the land, to how visitors and staff circulate, to requirements for individual spaces within the building (Figure 2.8). During the workshops participants will be asked to employ various techniques including writing and sketching so desks or seminar tables are important.

使用者

经常有这样的情况,将来的使用者包括访客同实际上的业主并不是一群人(业主是指委托建筑师设计项目并对其进行经营活动的实体),建筑师也无法找到足够多的使用者进行对话。这样的例子包括博物馆将来的参观者或是医院未来的病人。如果这些信息无法获知,参会的人员也无法回答,那么由甲方出面组织一个访客或是未来使用者的人员分析就是十分重要的。这样的研究可以提供一个对于未来建筑使用人口的基本认知,回答诸如"谁"、"多少人",以及他们的年龄、受教育程度等一系列重要的问题,这些都会对设计师产生巨大的帮助。在医院的例子中,对于未来病人人员构成方面的理解是初期项目可行性分析中十分必要的一部分,应该是现成可得的资料。在博物馆的例子中,因为设计是由藏品和资金驱动的,未来参观者人员的构成并不那么清晰。如果需要,这一研究将会是一个长期的过程,可能需要花费些时间才能完成。

可获得的文件

在场地分析(设计任务设定中间的一个关键环节)的准备过程中,建筑师将会要求查阅大量关于场地的文件,其中包括现有的规划分区规定和各种不同的勘测报告,包括场地的边界、地役权、现存的植被、地形(参看66页典型的清单)、土壤特征以及水文条件。再一次重申,如果这

Users

Often users include visitors who differ from the actual client (in the sense that the client commissions the architect and is the operating entity) and may not be available for the architect to talk to in meaningful numbers. This is the case, for example, with visitors to museums or patients in a hospital. It is important, if the information is not available, or known to workshop participants, for the client to commission a **visitor or user analysis**. This study provides demographic understanding particularly answering the "who", and "how many" along with age, education levels, and myriad important facts that help inform the designer. In the case of hospitals, an understanding of the patient demographic was necessarily part of preliminary feasibility study information and should be readily available. In the case of a museum, which is collection and funding driven, the answer might be less obvious. If necessary, this analysis is a long lead item and will take a significant amount of time to accomplish.

Available Documentation

In preparation for doing a site analysis, a critical component of the program, the architect will require access to extensive documentation regarding the site including current zoning, a title report and various surveys showing boundaries, easements, existing vegetation, topography (See typical list pg. 66), soils characteristics and hydrology. Once again, these items, if not already available, must be commissioned immediately to avoid causing delays. It should be noted that the architect requests that the owner provides this information,

些信息不是现成可得的，那么就应该立即要求业主安排进行调研以免造成项目的延误。有一点需要特别提醒的是，建筑师会要求业主提供这些信息，而不是自己去试图获得这些信息以降低法务上的风险，因为在很多情况下信息的有效性可能会超出了他们的控制范畴，例如实际的项目红线的位置或是一个特定的地役权的位置。

项目进度表

一般来讲项目决策人对于项目何时启用都会有一个清晰的概念。这个日期可能会由很多不同的因素来决定，例如财政资助的截止日期、政治上的需要、实际操作方面的考虑或是历史事件发生的时间。这个进度表可能会提前几年就设定好了，这中间并没有考虑到可能会造成项目延迟的意外情况。项目要依靠建筑师的理解来确定达到各个分段目标所需的设计时间，即设计任务的制定，由概念至施工图阶段的图纸设计（包含各主要阶段节点的确认）、雇佣施工团队、获得项目施工许可、土建施工、内部精装施工以及最终的室内配饰软装完成等各个阶段。通常来讲，资金筹措会由于其自身的一些限制条件成为项目进程关键的推动力。

预算

在一种完美的情况下，预算是根据设计任务中所列出的标准来决定的。但现实情况却是，在所知有限且提前数年的时候，项目会设定出一个预算用来讨论；拨款或是其他来源可使用的资金

---

rather than obtaining it themselves, to mitigate legal liability for situations that they have no control over such as the position of the actual property line or location of a particular easement.

Project Schedule

The DM will normally have a strong idea about when a building should be ready to occupy. This date may be driven by a financial deadline, political realities, practical necessity or an historical occurrence. The schedule may have been set years in advance with no contingency for the inevitable delays that preceded this meeting. It is up to the architect to understand and convey the realities about how long it will take to achieve major milestones, i.e. programming, completion of design through contract documents (including major approvals), hiring of a contractor, permitting, construction and interior fit out. Often, fundraising or financing will drive the critical path with its own set of contingent restrictions.

Budget

In a perfect situation, the budget would be determined based upon the criteria documented in the program.

中似乎是随意决定的一部分会被划分给建筑项目。预算信息可能是公开的，但项目前期的研讨会仍然是一个合适的场合来说明设计任务应该成为制定符合现有资金量的预算标准的工具（尽管可能不能够立即满足所有的需求）。

尽管我们对这个问题讨论得不多，但我们需要记住，对于项目决策人而言，特别是在一些政府机构中，预算和进度计划对于项目的成功而言往往是至关重要。

项目决策人的视野

无论是在会议的开始，还是在研讨的结尾或是在一场单独的茶歇聚会中，与项目决策人的沟通都很关键，不管是被用什么样的方式告知，诗意的或是刻板的，建筑师都希望了解到他们对于项目的愿景是什么。我们为什么要创造这个项目？什么是关键的因素？怎样去定义成功？实施计划又是怎样的？

项目决策人会在很大程度上影响项目的走向，这一风险会贯穿项目的始终，从挑选建筑师开始直至设计工作的完成。如果，按照正常的情况，项目决策人选择让建筑师依据设计任务中的需要去确定建筑的形式或外观，这就意味着建筑师需要让尽可能多的利益相关者参与到项目讨论中来。但是这里需要说明的是，也许让一些人（员工、公众）自由地说出对于项目需求的想法并不是

The reality however, is that often years earlier, based upon limited knowledge, a budget was set for discussion purposes; an appropriation or other funding was put into play and an often seemingly arbitrary number was divined for the building. While budget information may be public, the workshop is an appropriate place to explain that the program should be the tool to establish criteria for scenarios (which may not immediately meet all needs and wants) that can meet available monies.

Although there will be little discussion here, it is important to remember that to the DM, especially in a bureaucracy, the budget and project schedule are often the most crucial criteria for success.

Decision Maker's Vision

Whether at the beginning of the meeting with the Decision Maker (DM), at the end of the session or in a separate get-together over drinks or coffee, it is crucial for the architect to be told, in terms either poetic or pragmatic, this person's vision for the undertaking. Why are we creating this project? What are the critical issues? What defines success? What is the implementation plan?

The DM will strongly influence the direction that this venture takes commencing with the selection of the architect through the completion of design. If, as it should be, the DM leaves the architect to interpret the building form/appearance based upon the program requirements, it remains for the architect to gather input

一件容易的事。如果建筑师观察到有些人在大型的会议上沉默寡言，那就应该想办法在一些更放松的场合鼓励他们表达出自己的意见。在任何情况下，对于建筑各部门或是不同系统如何运转的问题，同各部门的领导或总监一对一面对面地沟通都是建筑师获得专家意见的最佳途径。经常的，就如之前提到的那样，想法上的碰撞（互相矛盾的元素）可以引领建筑师找到最佳的解决方案。

需要注意到的是，有时候项目决策人所信奉的哲学可能同他所领导的机构中的其他人的意见有冲突。当建筑师将工作向前推进时，据此所进行的设计会带来严重的问题。在一个有趣的案例中，一个机构的董事会不同意动物园总监关于在非常狭小的空间里展示大型活体哺乳动物的提议。博物馆的总监，作为这个项目的负责人（决策人），觉得一些小型的哺乳动物，例如啮齿类动物可以被安置在一个合适于它们那种尺度的小一些的生存环境里，通过对它们的观察来有效地向研究它们习性的人开展环境方面的课程。鉴于之前提到的矛盾和这样的思路，他将一些大型的哺乳动物例如狮子从展示清单中移除了。最终，这名项目负责人，由于公众的抱怨而被解除了职务，毕竟公众来到这里参观就是希望能看到一些令人惊叹的野兽的。

在另一个有趣的例子中，一个富有远见的国际法律部门的总监传递给建筑师的信息是他相信他们部门的成功是基于在一个开放的环境中，员工之间有自由地交流和互动。而另一方面他手下的律师团队却不同意这个观点。在同建筑师的单独交流中，他们一致认为唯一可以让他们高效完成工作的方法是给他们创造出一个私密和安静的空间。根据这一组看似互相矛盾的信息，建筑师得以创造出一个独特的解决方案，既满足了总监关于互动交流的设想，又照顾到了他的团队渴望相对独立的工作环境的要求。

---

from as many stakeholders as possible. It should be noted, however, it may be difficult to get others (staff, public) to speak freely about their ideas regarding project needs and wants. If the architect observes that individuals are reticent to speak in large meetings, they should be encouraged to provide information in a more comfortable setting. In any case, one on one meetings with various directors and department heads are essential for the architect to get expert input on how each part of a building or system should function. Often, as mentioned before, competing ideas (contradictions) can lead the architect to the best solutions.

Note that the philosophy which molds the DM's vision may be in conflict with others involved in his organization. This can lead to severe problems with the basis of design as the architect moves forward. In one interesting case, the Board of an institution disagreed with a zoo's director over the display of large, live mammals in very confined spaces. The museum director, the Decision Maker, felt that small mammals such as rodents, could be housed in habitats of appropriate scale to their size and through their observation effectively teach important environmental lessons to those who studied behavior. Accordingly, he removed certain large mammals, such as a lion, from display. Ultimately, the DM, due to a grumbling public, who, after all, came to see a magnificent beast, regardless of the pain of its confinement, was removed from his post.

In another interesting case, a visionary director of an international law think tank conveyed to the architect that he believed his institute's success was based upon the free flowing interaction and exchange of ideas among his staff in an open environment. His team of attorneys on the other hand disagreed. In private discussions with the architect they uniformly stated that the only way they could effectively accomplish their tasks was to be in private, quiet space. From this seemingly contradictory information the architect was able to create a solution that met the director's vision for interaction while enabling his team to work in the solitude they desired.

在同项目决策人会面后
详细的进度表

　　有两种与计划制定设计任务相关的进度表。第一个是之前提到的整体性的进度表，项目负责人同建筑师一起确定设计任务讨论会的时间。这就为下面讨论到的更详细的进度安排确定了一个起始时间。

　　第二种类型的进度表，是在同项目决策人会谈之后，对于制作出设计任务文件所需要完成的工作的一个更详细的梳理。建筑师将所有与制定设计任务相关的活动都列出来，包括宣传、联系场地、落实会议主持人、设定相关会议、场地考察、相关专题的分析研究以及最终的设计任务文本制作等（参看附录A.2）。以我们的经验，对于制定设计任务相关的所有活动的最佳组织方式是使用一个叫作关键路径进度表的工具，这是一种将各个团队活动以及它们所需的时间用一种逻辑网络模式链接在一起直至最终的工作完成的图表（参看附录A.1）。

　　这种类型的进度表的有用之处，除了可以标示出每一项活动所需花费的时间之外，还可以在最早的时间就识别出长期的活动是什么。长期活动是指那些通常要花费较长时间才能完成的事情。它们应该在很早就被计划安排好以免成为延误最终结果的因素。特别是这样的活动同收集信息相关时，它们可能包括制定或是细化任务宣言，完成环境影响因素报告等。可以很容易地看出如果不在最早的时间布置需要的专题研究或是勘察会对项目造成多么严重的延误（图2.10）。

AFTER MEETING WITH THE DECISION MAKER
Schedule - Detailed

There are two types of schedules associated with planning the programming effort. The first is the generalized schedule when as mentioned earlier, the DM and the architect block out time for programming meetings. This establishes a start date for the more detailed scheduling discussed following.

The second type of schedule, created after meeting with the Decision Maker, is a detailed look at what needs to be accomplished to produce the Program Document. The architect begins by listing all activities associated with making the program including publicity, obtaining venues, facilitator commitments, meetings, site visits, research and program production, etc (See appendix A.2). From our experience, the best way to organize all of the activities leading up to the creation of a program is to use a critical path schedule; a chart which links contingent activities and their durations in a logical network leading to the completion of the work (See appendix A.1).

A useful aspect of this type of schedule, in addition to establishing the duration to accomplish a set of activities, is identifying at the earliest moment, what are known as long-lead items. Long lead items are those things which normally take a long time to obtain. They should be scheduled early enough that their avail-

（图2.9）这是一张典型的项目行动逻辑关系图，上面标明了不同行动的时间、依托的条件以及每个行动完成时所要达成的目标。在节点任务（大圆内）之间，为了项目的成功，有一系列需要按照顺序完成的行动。

(Figure 2.9) A typical project network logic diagram charts activities and their dependencies, time, and end points called milestones. Between milestones (large circles), there are a number of different activities that must be performed in sequence in order for the project to succeed.

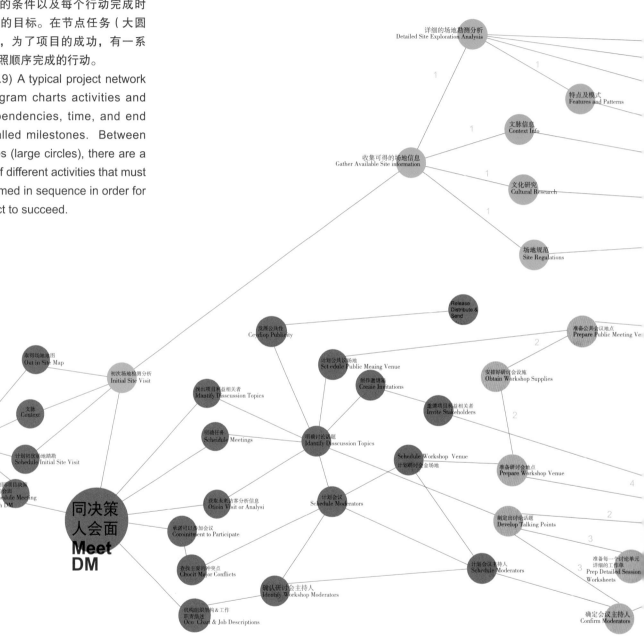

设计的开始
Beginning

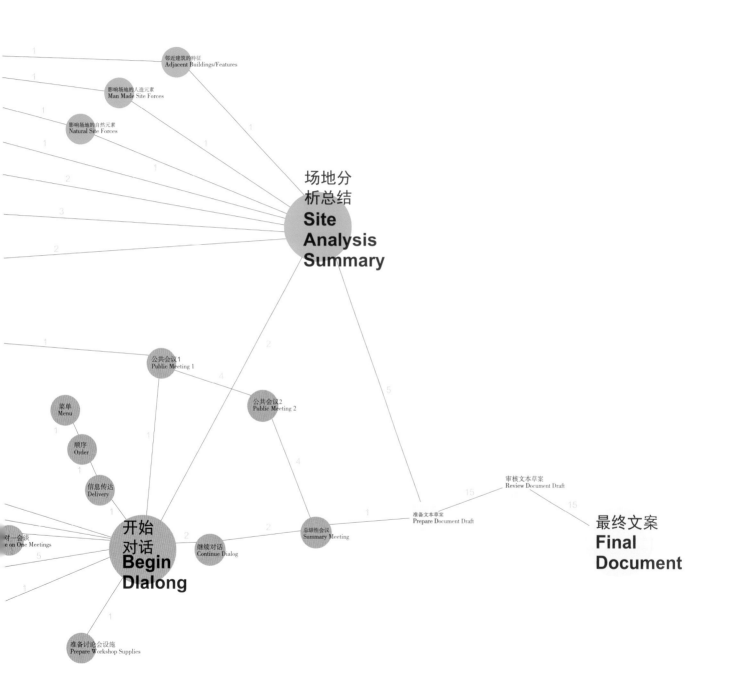

通常来讲，一些非常耗时的工作，例如环境影响分析报告（可能会花费1～2年）、交通分析以及访客分析等，一般在客户早期的项目可行性分析的过程中就已经完成了。

有趣的是，对于设计师理解场地和文脉非常有帮助的航拍照片，原先作为一项最昂贵且耗时最长的工作，现在可以很容易地即刻在谷歌地球上获得。这说明所谓的长期活动在不断地被重新定义中，这取决于我们所依赖的科技、可咨询到的专业人士以及要求的范围有多大。

---

ability or completion is not a factor in delaying the end result. As they specifically relate to gathering information, they could include development or refinement of the mission statement, completing environmental impact narratives, etc. It is easy to see how not commissioning a needed study or survey at the earliest time can lead to significant project delays (Figure 2.10).

Normally, very long lead items such as environmental impact statements (1 – 2 years), transportation studies and visitor analysis will have been completed by the client at an earlier time as part of a project – go, no-go -- decision making (feasibility) process.

Interestingly, one of the most expensive and time-consuming long lead items, the engaging of aerial photography to advance both contextual and site understanding, is now instantaneously available on Google Earth. This points out that long lead items are constantly being redefined depending upon technology and availability of expertise as well as scope of the requirement.

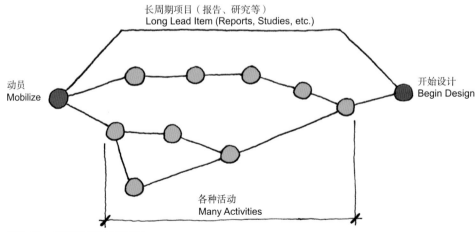

（图2.10）注：一项活动完成所需要的时间长度等同于其他很多活动完成时间的总和，这样的活动我们称之为长期活动。
(Figure 2.10) Note: A single activity taking as long as the aggregate total of many other activities is a long lead item (activity).

图2.10　长周期项目示意图
Figure 2.10 Long Lead Item Diagram

Typical Agenda Checklist

Mobilize
- **Schedule meeting with Decision Maker**
- **Schedule site visit**

People
- **Decision Maker commitment to participate**
- **Identify key participants with contact info.**
  - Stakeholders
    - Staff
      - Org charts
      - Job descriptions
      - HR -- Policy manuals
    - Board
    - Directors/Department heads
    - Staff
    - Politicians
      - National, State and Local
      - Staff
      - Constituents
    - Regulators
    - Public
      - Supporters
      - Naysayers

Strategic Plan
- **Verify Mission Statement**
- **Decision Maker's Vision – why (not how)**
- **First this than that**
- **Identify Tactics and Strategies**

Construction Budget
- **Determined by available funds**
  - Appropriation
  - Bond issue
  - Taxing district – revenue stream
  - Gift funding

典型的议程表清单

动员
- 同决策者商定会议日程
- 安排实地考察日程

人
- 决策者承诺参加会议
- 确定主要参与者和联系人信息
  - 利益相关者
    - 员工
      - 组织结构图
      - 职位描述
      - HR – 政策手册
    - 董事会
    - 总监或部门主管
    - 员工
    - 政客
      - 国家、州和地方
      - 员工
      - 成分
    - 监管机构
    - 公开
      - 支持者
      - 反对者

战略计划
- 验证任务描述
- 决策者的愿景——为什么（而不是如何）
- 优先要解决的问题
- 确定战略与策略

施工预算
- 通过现有资金确定
  - 拨款
  - 债券发行
  - 征税地区——现金流
  - 赞助款项

Program Schedule
- **Availability of key participants**
  - Decision maker
  - Board
  - Public
- **Key community date conflicts**
  - Community/National events
  - National Holidays
  - Religious Holidays

Framework for Gathering Information
- Format
  - Workshops and Meetings
    - Formal or informal
    - Duration
    - Where (classroom, on site)
    - General scheduling

Available Information
- Site Data
  - Owner furnished
    - Survey
      - boundary
      - topographical
    - Title
    - Zoning
    - Existing plans
  - Publicly available
    - Photography
    - Google Earth
    - Utilities
    - Access
    - Easements

Possible Long Lead items
- Visitor Analysis
  - Numbers
    - Frequency

设计任务讨论会日程安排
- 关键参与者的档期
  - 决策者
  - 董事会
  - 公众
- 同社区日程的关键矛盾点
  - 社区或国家事件
  - 国家法定节假日
  - 宗教节日

信息收集框架
- 格式
  - 研讨会和会议
    - 正式或非正式的
    - 长短
    - 地点（教室、在现场）
    - 总体日程安排

可获得的信息
- 场地数据
  - 业主提供
    - 调研
      - 边界
      - 地形
    - 信息名称
    - 功能分区
    - 现有规划
  - 信息对公众的开放度
    - 照片
    - 谷歌地球
    - 公用设施
    - 场地通达性
    - 地役权

可能的长周期项目
- 访客分析
  - 数量
    - 访问频次

设计的开始
Beginning

- Calendar
- Demographics
  - Ages
  - Education
  - Income
  - Domicile
- Transportation model
- Desired services
- Critical issues
- **Environmental Studies**
  - Assessment
  - Impact Statement

Identify Long lead Items
- **Reports and Studies**
  - Hydrology report
  - Soils report
  - Archeological studies
  - Environmental studies
    - Environmental Assessment
    - Environmental Impact
  - Survey

- 日历
- 人口成分
  - 年龄
  - 教育程度
  - 收入
  - 住所
- 交通模式
- 所需服务
- 关键问题
- 环境研究
  - 评估
  - 影响报告书

识别长周期项目
- 报告和研究
  - 水文报告
  - 土壤报告
  - 考古研究
  - 环境研究
    - 环境评估
    - 环境影响
  - 调研

场地
Place

建筑建造的地点；它关注的是具体位置的细节特征。

Where architecture is made; it concerns the specifics of location.

场地所关注的是建造场所的具体情况，以及建筑将要被建造的位置的地质条件。如果我们希望能够同自然及人文环境和谐相处，那么如何恰当应对建筑所处的场地就成为设计成败的关键。除了一些客观条件例如场地特殊的气候、高程、地质、水文、生物及动物等因素外，我们所说的场地还包括了具体建造位置同其所处的自然环境和更大的人文环境间的关系；人文因素可能包括了景观视线、噪音、社区、地区、城市、省或州、国家甚至是地球（图3.1）。在这一章节中我们会关注从一般性的场地影响因素具体到实际建造场地的环境和文脉。

场地影响因素

如果，像之前所说的，形式追随知识（功能），那么就会很容易地看到对于场地的研究可以在多大的程度上影响到最终的建筑设计。当遇到一个陡坡，建筑应该是阶梯状的顺坡起势，埋进土里，或是吊脚楼似的立于地形之上？是否我们通过太阳运动的路径加上一年中晴朗的天数就可以决定玻璃安置的方式以及阳光及能源是应该被引入室内还是该被阻隔于室外？我们能够导入并重新定向夏天凉爽的微风来作为空调系统的一种低成本的替代方案吗？对于建筑来说，我们的设计是否会加快风速或是改变风向？如果是这样的话，会有什么样的影响呢？我们是否可以利用昼夜间的温差让居住者可以几乎是零成本的去控制室内环境的舒适度？

规划条例和规范只意味着乏味的雷同性还是可能为创新提供出一条不一样的途径？一个这样的例子可以在下面这张图表中被看到（图 3.2）：在这里，我们设计的在沙漠环境中的室外剧场既要满足美国残疾人法案关于轮椅可达性的要求，又需要创造出阴凉。在剧场内设置坡道使得轮椅可以到达设计任务要求的最边远的看台区域。因

Place concerns the particulars of the locale; the geography, where architecture will be made. If we hope to live in harmony with our planet responding to forces, both natural and man made, acting upon the specific building location is essential. In addition to the realities of site specific climate, topography, geology, hydrology, biology and zoology, place includes the location's relationship with natural and man made forces of the greater context; views, noise, the neighborhood, district, city, province/state, country and even the Earth (Figure 3.1). In this chapter we will be considering both site influences specific to the actual building site and context.

Site Influences

If, as stated earlier, *form follows knowledge*, then there are numerous clues about how a building should be gained from studying the place. Presented with steep terrain, does the building step with the slope, burrow into it, or rise above it? Does the path of the sun coupled with the number of clear days dictate how glass will be placed and energy and light captured or repelled? Can we scoop, deflect, redirect or otherwise capture summer breezes that offer a cooling, low cost, alternative to HVAC? By building, do we cause the wind to speed up or change direction? If so, what are the implications? How about the role of high diurnal temperature changes in allowing occupants to take control of their comfort at nearly no cost?

Do zoning regulations and codes dictate a dull sameness or can they offer a different route to creativity? An example of this can be seen in (Figure 3.2), where the design for a desert amphitheater became a relevant piece of architecture as a result of complying with both the American Disabilities Act's (ADA) wheel chair access requirements and the need for shade. Locating an incline within the amphitheater to allow wheel chair access required area far in excess of the programmed seating requirement. Hence, the opportunity was presented to plant deciduous trees in planters inside the amphitheater to shade the attendees in summer. Can

场地
Place   71

（图3.1）谷歌地球是一个强有力的工具，它能够帮助建筑师探寻项目场地的尺度和内容。左边是三张不同尺度的地图，最上面一张红线区域内是场地范围，下面的两张则显示出更大区域范围内的文脉关系。

地图数据：

谷 歌，DigitalGlobe, Europa Technologies, Mapabc.com

(Figure 3.1) Google Earth is a powerful tool to help the architect explore the project site context and scale. The series of three maps at left shows a site's property line in red at a scale specific to the site (top) and progressing to more general or context focused at the bottom.

Map data: Google, DigitalGlobe, Europa Technologies, Mapabc.com

此，我们有机会通过在剧场中的花池里种植落叶树在夏天为观众提供阴凉。我们是否可以通过和景观设计的无缝连接将一部分室内功能移至室外，从而减少需要空调的室内空间？

有了这样的认识，并且试图去找寻这类问题的答案，我们就会很容易地看到场地本身如何去影响建筑的设计。作为建筑师和教育家，美国建筑师协会院士团成员理查德·威廉姆斯喜欢说的一句话是"the Genus Loci（场所的精神）引领我们的思考"。我们通过土地和相关利益者的需求（参看下一章节——对话）共同决定出的概念慢慢地展开建筑设计。

设计信息的收集包括了去发现建造场地周围环境对建筑的影响。场地研究最基础的部分就是探索和观察，去到现场走一走，看一看。作为项目建筑师，必须要参与到下属团队成员的行动中才能最大化地发挥这一工作对于建筑设计的影响。而这一工作的基础就是利用照片以及在基准地图上的标注记录下场地调研时所观察到的信息。

一份好的基准地图一般会标示出项目边界及高程测绘信息、现场植被、邻近的结构物、道路以及其他交通系统。由于野外风的影响，A3大小的地图是最适合掌控的尺寸。另外可以多准备几份拷贝用夹子夹在风波板（KT板）上方便标注信息，不同的拷贝上可以专门用来记录一个或两个主题的信息，例如景观视线或是噪音等级，这样的安排可以保证我们有足够的空间对信息进行记录（图3.3）。为了使基准地图和实际的场地吻合，在设计师前往场地考察之前，业主所雇佣的测绘人员必须将场地的边界及转角处清晰地标示出来（最大间距每隔30m用1.2m高的旗子标出）。

一张或多张基准地图将会被用来记录同照片相关的信息。我们推荐的方法是在一个固定点上照多张照片最后拼成360°的全景照片，这些位置被称为照片拍摄点（图3.4），并分布在整个场地中。另外我们还会补充一些单独的照片以记录特殊的植被、噪声来源等信息。如果可以将这些照片拍摄位置标注到基准测绘地图上则是最理想的。

---

the outside of a building merge with the landscape to offer seamlessly connected exterior places of contemplation and collaboration while resulting in reduction in air conditioned interior space?

With this in mind, and by developing answers to these sorts of questions, it is easy to see how the place itself begins to influence the design of the building. As the architect and educator Richard Williams, FAIA, is fond of saying "the *Genus Loci* (spirit of the place) guides us in our thinking". As the concept is informed by both the land and stakeholders needs (See dialog chapter), we begin to see the architecture unfold.

Being *informed* includes discovering the attributes of a building site's surroundings. The most fundamental aspects of site understanding are exploration and observation; the actual walking, and seeing the place. For maximum architectural influence, the design architect must participate along with members of his team. Fundamental to this investigation is recording observations through photography and notation on a base map.

A good base map will generally indicate boundary/ topographic survey information, vegetation, adjacent structures, roads and other transportation systems. This map is most manageable for fieldwork – due to wind - when A3 (11" x 17") sized. Multiple copies should be clipped to foam core board for ease of note taking and may be used individually for one or two topics such as views and noise in order to have sufficient room for notes (Figure 3.3). In order to assure that the base map and actual site are synchronized, the property corners and boundary (staked at a maximum of 30M on center with flagged lathe 1.2m – 48" tall) must be well marked ahead of the site visit by the owner's surveyor.

（图3.2）一些强制性的规范条例可能会导致一些有趣的设计方案出现。下图就是一个为了满足遮阴、无障碍规范以及雨水收集等需求而设计的室外剧场空间。

(Figure 3.2) Code requirements may lead to interesting design solutions. Amphitheater responding to the need for shade, ADA Accessibility, and water harvesting requirements.

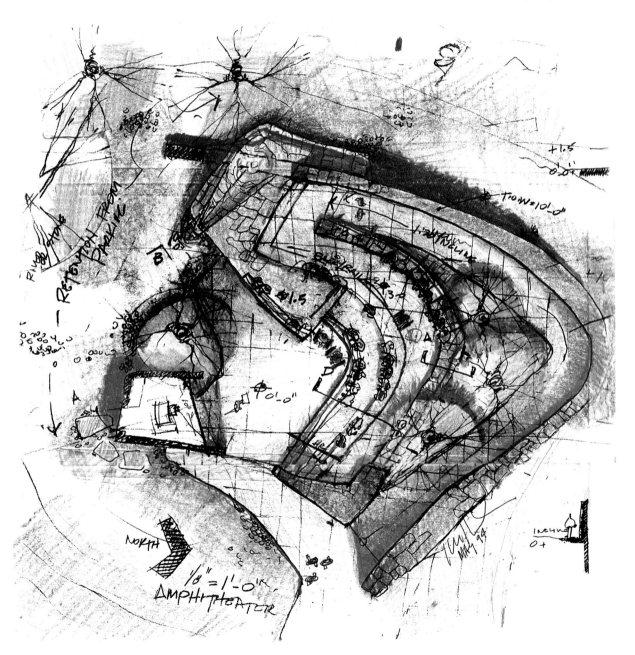

在开展设计工作之前，一些技术性的资料整理工作必须要完成（由专业测绘人员完成），这包括以下一些内容：例如定位场地的边界以及临近道路的情况、退线要求、合适尺度的高程信息（开始时可以是1m高差间隔的等高线，最终达到最小300mm间隔的精细程度）、洪泛区域的边界、场地排水方式、特异的地质景观（如突出的岩石或是裸露的大石块）、植被信息（树干直径大于10mm或是更大尺寸的枝干组团）包括生长的方式、地役权方面的信息（特别是关于市政管线、机动车以及行人的通行权利），自然的和人为的现有条件例如现有道路、小径、围栏、墙体、构筑物，所有市政管网设施的位置及容量包括通信设施（电缆、光纤、电话线）、电力、天然气、水、污水和雨水的排放管道，以及私人和政府间的法律规范例如留置权。

由设计者的观察和政府历史资料提供的场地自然影响因素（有关于飓风、洪水和地震的历史数据、100年来最高最低和平均气温、最多最少和平均降水以及不同季节的主导风向与风速）对于建筑设计来说是一些最基本的信息准备（图3.5）。同理，不同场地的日照角度变化也直接影响着设计师的思考。

在最近的几年中，气候的变化很明显地比近代历史上的任何时期都要快。特别需要提到的是，似乎风暴发生的强度和频率都在增加，沿海处的海平面也在逐渐升高。纽约现代艺术博物馆最近举办了一个名为"洋流"的展览，希望引起人类对于可能同这样的现象共存的注意。尽管对于极端气候的解决方案已经远远超出了本书的范畴，但建筑师还是可以将这些因素作为场地分析的一部分加以关注。

One or more base maps will serve to record information related to photography. We recommend utilizing a number of 360 degree panorama views from fixed locations, known as picture points (Figure 3.4). throughout the site. These are supported with individual photographs of vegetation, noise sources, etc. It is best if the picture points are staked and tied into the maps by survey.

Technical documentation (by surveyors), necessary before design can commence includes locating site boundaries and adjacent right of ways along with property line dimensions and bearings, setbacks, topography at appropriate scale - initially a 1M (3'0")interval and eventually a minimum contour interval of 300mm (1'0"), flood plain boundaries, drainage patterns, geological anomalies such as outcrops or large boulders, vegetation (trees with diameter of 10mm (4") and larger and dominant clumps and patches) including patterns of growth, easements (particularly for utilities, vehicular and pedestrian passage), natural and man made features such as existing roads, paths, fences, walls, structures and location and capacity of all utilities including communication (cable, fiber optics, telephone line) power, gas, water, sewer and storm drainage pipes, as well as private and governmental encumbrances such as liens.

Natural forces determined by observation and researching government records (historic data regarding hurricanes, floods, and earthquakes as well as 100 year high, low and average temperatures, maximum, minimum and average rainfall, flood records and maximum seasonal wind direction/speed) are fundamental to informing the building concept (Figure 3.5). In the same way, the ever moving sun angle directs the designer.

In the last few years it has become apparent that weather patterns are changing more rapidly than in recent history. In particular, it seems that storm intensity and frequency is increasing and coastal waters are rising. A recent exhibition, *Currents*, at the Museum of Modern Art in New York City calls our attention to possibilities for coexisting with this phenomenon. While solutions are well beyond the scope of this book, architects are particularly well suited to address these factors as part of their site analysis.

场地
Place

（图3.4）用文案的方式记录下影响场地的因素，在这个例子中是景观视线，这些因素会成为决定建筑将来设计成什么样子的基础。
(Figure 3.4) Documenting site forces, in this case views, is a fundamental aspect of defining what the architecture will become.

近景 Near view
中景 Intermediate view
远景 Distant view
被前景遮挡的远景 Distant view obstructed by foreground
视线 Lines of sight
景观延伸 Extent of view

1 茂密植被景观 View of dense vegetation
2 灌木景观 View of scrubby vegetation
3 图森山脉景观，近处的山峰 View of Tucson Mts. near peaks
4 鸟舍景观 View of aviary
5 展示花园景观 View of Demonstration Garden
6 后勤院落景观 View of maintenance yeard
7 图森山脉景观，远处的山峰 View of Tucson Mts.far paks
8 山谷及远处山脉的壮丽景色 Spectacular view of valley and distant mountains
9 山谷景观 View of valley
10 被展示花园遮挡的远处的景观 Distant view obstructed by Demo. Gardens
11 被图森山脉近处的山峰遮挡的远处的景观 Distant view obstructed by near peaks of Tucson Mts
12 被鸟舍遮挡的远处的景观 Distant view obstructed by aviary

（图3.3）为了能在和项目利益相关者讨论时抓住关键因素，在现场考察时做认真详细的笔记非常重要。
(Figure 3.3) Taking detailed notes about observations while onsite is essential in order to discuss critical issues with stakeholders.

（图3.5）包含有天气资料例如平均气温、降水以及风向的气象图表可以从国家气象部门获得。
(Figure 3.5) Climate charts for average temperature, rainfall, and wind patters are available from National Athmospheric Agencies.

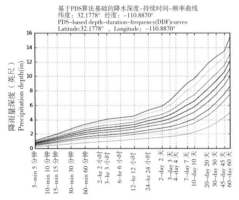

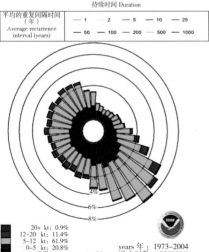

文档VS. 分析

场地文案记录和场地分析之间是有区别的。场地文案记录了那些与场地有关的事实，一般最终会以书面的形式记录在一个地图上。而场地分析则要求设计师通过用眼睛观察这些记录，进而解释这些事实会怎样去影响未来的设计。举例来说，如果记录到有洪水发生，那么场地分析就需要告诉我们这个场地是否可用，降低这种自然危害的策略有哪些（图3.6）。

当我们在对中国重庆的一个高尔夫球会所项目的场地进行勘察记录时，我们在场地中发现了一个废弃的矿坑。尽管开始的时候大家认为这是一个麻烦，但经过场地分析后我们却发现它其实是一个财富。临近的道路计划下挖6m深——设想怎样从公共道路连接到建筑。这个问题，连同前面提到的利用废弃的矿坑创造出一个有趣的入口庭院的想法，开始从文化思想及物理形态两个方面暗示出设计该如何进行。我们的场地记录（图3.7）显示矿坑中的积水液面同计划下挖的道路标高持平。按照正常的逻辑推理，停车场应该下挖到同道路相同的标高。这样的话，停车场仍然可以直接联系到入口庭

Documentation VS. Analysis

There is a difference between site documentation and site analysis. Site documentation records the facts having to do with the site, generally summarized in final form, on a single map. Site analysis looks at the facts (documentation) with an eye to explaining how they will affect the design. For example, if flooding is noted, analysis dictates that we determine if the site is usable and what the reasonable mitigation strategies may be (Figure 3.6).

While documenting attributes of a golf clubhouse site near Chongqing, China, we noted an abandoned quarry. While initially considered a hazard, analysis suggested the quarry was an asset. A nearby planned road cut - 6m deep - posits how to connect from this road to the building. This question, along with the idea that the quarry will make an interesting entry court begins to inform the design culturally as well as physically. Site documentation (Figure 3.7) shows that the water level in the quarry is at the same level as the planned road cut. Logic indicates that required parking

院。经过分析，我们提出了在现有的砂岩上切割出一条狭窄通道的方案。这条狭窄的通道会先压抑一下参观者的情绪，当参观者通过后到达入口庭院时就会豁然开朗，这远胜于直接开门进入建筑空间（图3.8）。

should be cut to the same level as the adjacent road. Now, all that remains is to connect from the parking to the entry court, and through analysis, we proposed a narrow passageway cut into the sandstone. This passageway would serve as a compression element so that when the visitor reached the entry court it would seem even more expansive than if approached directly (Figure 3.8).

（图3.6）一些看起来场地中的负面因素，例如这条河床正好穿过住宅场地，其实可以被巧妙地融入设计构思从而创造出令人兴奋的建筑。这张照片中，工人正在把一个预应力梁做成的跨桥搭建在河床之上，跨桥将分置业在河床两侧的住宅中的公共和私密区域连接在了一起。

(Figure 3.6) Seemingly negative site features, such as this arroyo that runs through a residential site, can be incorporated into designs to create exciting architecture. Here, a bridge is placed by construction workers that will span the arroyo and connect between the public and private areas of the residence.

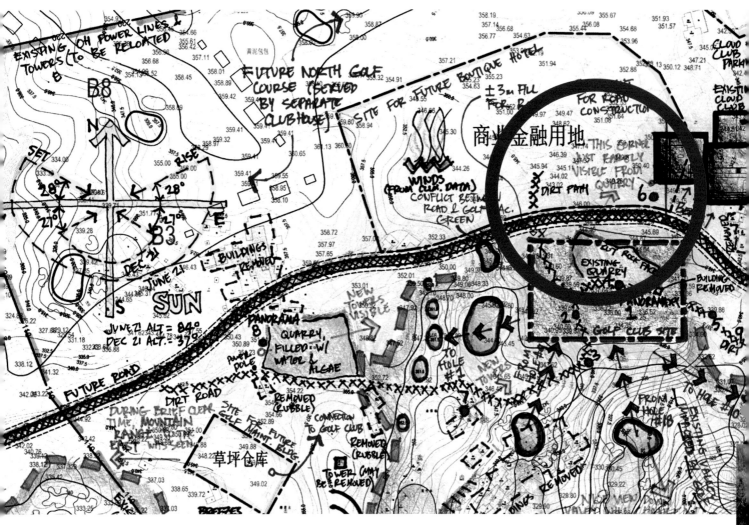

（图3.7）这张场地信息记录示意图上标示出了一个现存的石矿（用红色线圈出的地方），它作为场地中的资源被善加利用并影响到了最终的设计。(Figure 3.7) The site documentation diagram identifies the existing quarry (circled in red), as an asset that may be used to inform the design.

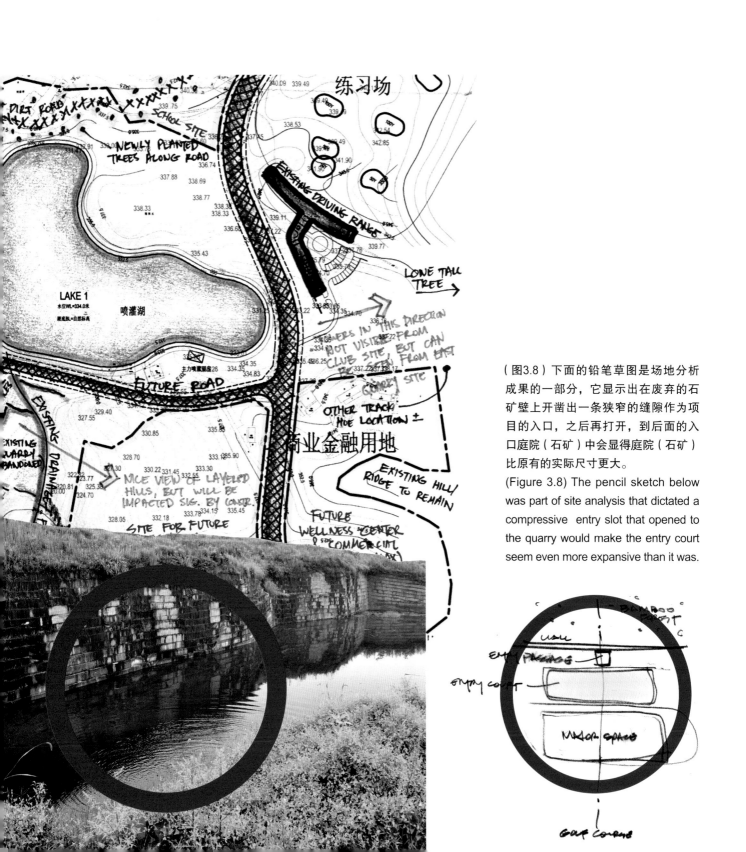

（图3.8）下面的铅笔草图是场地分析成果的一部分，它显示出在废弃的石矿壁上开凿出一条狭窄的缝隙作为项目的入口，之后再打开，到后面的入口庭院（石矿）中会显得庭院（石矿）比原有的实际尺寸更大。

(Figure 3.8) The pencil sketch below was part of site analysis that dictated a compressive entry slot that opened to the quarry would make the entry court seem even more expansive than it was.

理解文脉

当建筑师谈到文脉，他们关注的是场地边界以外更大的区域将会对一个特定建筑所产生的影响。很明显，当附近计划修建高速公路或是架设一条新的高压传输线的时候，这种影响是显而易见的，但建筑师所需要考虑的远不止这些，如何尊重当地的风土人情，如何融入更大的环境并同邻居们和谐共处对于建筑师来讲是非常重要的。正如前面章节"意象"中（参看32页）所讨论的那样，这并不意味着建筑要去复制周边其他建筑的样式或是参考某种特定的历史风格。如前所述，建筑设计核心的一点就是要忠实于它的时代和场地。

对于文脉理解的过程同对于场地影响的理解类似。它是基于严谨的研究以及落到实处的对于一个场地的探索、观察、测量、记录和体验。周边的结构有多大？多高？它们的阴影在哪儿？我们的阴影呢？人们都会从什么地方来到这里？怎么来？我们怎么和邻里产生互动？关于尺度，借助现有的科技手段例如谷歌地球（图3.1），我们可以很容易地立即掌握建筑、交通系统以及开放场地之间的相对尺寸及远近关系。在研究完手边的平面资料后（街区地图、旅游地图或是工程地图），设计师还必须亲自去现场考察。

下面是建筑师在试图全面了解场地文脉时会需要回答的诸多问题中的一部分。

建筑场地周边的区域是郊区还是城市？各个方向上政府的功能规划是什么？有什么周边的发

Understanding Context

When architects speak of context they are considering the larger setting outside the site boundary that will affect a specific building. Obviously, this matters greatly if a new power transmission line or highway is planned nearby but, more than that, it is important for architecture to fit well in its *genius locci*; to be a good neighbor and belong as part of a larger family. This does not mean, as you have already seen in the discussion of *Imagery* (See programming chapter pg. 32) that a building must be a copy of others or even a reference to some historical style. As stated before, it is fundamental that architecture is of its time and place.

The process of understanding context is similar to understanding the site influences. It is based in research and grounded in actual exploration; observing, measuring, documenting and experiencing a place. How big are surrounding structures? How high? Where are their shadow? Our shadow? Where do people come from and how do they get here? How do we interact with our neighbors? Scale matters, and now with technology such as of Google Earth (Figure 3.1) it is easy to immediately grasp the relative size and proximity of buildings, transportation and open space to one another. After studying available plans (street maps, tourist maps or engineering maps), the designer must personally explore the area.

Following are some of the many questions that the architect answers to become fully informed about the context.

Is the area surrounding the building site rural or urban? What is the government zoning in all directions?

展计划包括交通会对项目产生影响吗？周边的人口构成是什么？周边可以利用的公共（出租车、公交大巴、有轨电车、地铁、火车和飞机）以及私人（小汽车、飞机）交通有哪些？场地距离本地车站和主要交通枢纽有多远？哪些道路可以被看作是交通干道？附近有没有什么道路的修整、扩大或是移除的计划？旁边的建筑尺度是多大？旁边的土地形式以及主要植被或作物是什么？邻近社区的主导性建筑材料是什么？

当地的文化和传统如何去影响建筑的外观？是否利用本地人习惯从东边进入建筑的传统可以降低冬季北风的侵害？建筑场地周围的地质条件（排水、高程、气候和人口构成）如何？市政管网例如给水、天然气、污水、通信以及电力的容量、传输和分布的情况如何？有扩张的计划吗？公共安全部门（消防及警力）如何服务于场地？场地是否在一个"紧急通道"上？周边的医院和学校是什么类型的？在哪儿？场地中生活的人如何获得日常供给，例如加油、食品以及购买衣物？周边的一些餐饮娱乐设施，包括餐馆、博物馆、体育场或是艺术商店都在什么地方？各个方向的景观条件如何？距离工业设施有多远？受周边实体影响的环境质量如何？周边土地的价值高吗？附近地价的趋势是上升还是下降？怎样去筹款支持当地的学校、图书馆等？

我们可以通过观察、委托报告、手边可得的或委托进行的测绘或是利用一些公开发表的资料，例如地图或是人口普查的数据来得到这些问题的答案。同场地调研类似，对于文脉分析，

Are there neighborhood or regional plans in effect including transportation? What is the surrounding demographic? What is the availability of public (cabs, buses, trolleys, subways, trains and planes) and private (automobiles, planes) transportation and proximity of both local stops and major hubs? Where are the roads and highways? Which roads are considered arterials and are there plans for modifying, enlarging or eliminating roads and other modes of movement? What is the scale of nearby construction? Scale of nearby land forms and vegetation? What are the dominant building materials of neighbors?

How does culture and tradition affecting building appearance? Does utilizing the local Native Peoples' tradition of entering a structure from the east offer relief from prevailing winter winds? What is the geography (drainage, topography, climate and population) outside the building site? Where are and what is the capacity of transmission, and distribution of utilities such as water, gas, sewer, IT and power? Are there plans for expansion? How does public safety (fire and police) serve the place? Is the ground on an "emergency corridor"? Where are and what kind of hospitals and schools are nearby? Where does one obtain the necessities of everyday life such as fuel, food and clothing? Where are surrounding restaurants and entertainment including museums, sports venues and the arts? Where/what are the views? Where are the parks? Where does noise come from? What is the proximity of industry? How is environmental quality affected by other entities? What are surrounding land values? Are prices trending up or down? How are monies raised to support schools, libraries, etc.?

The answers to these and other question are obtained by observation, commissioned reports, available/commissioned surveys and utilizing published materials from maps of various scale to surveys such as

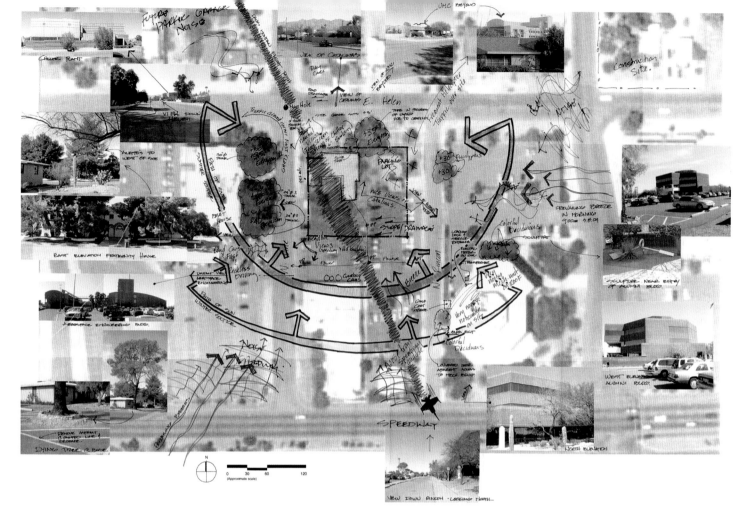

（图3.9）这是一张使用基准地图绘制的亚利桑那大学诗歌中心项目的场地文脉分析图，上面布满了各种关于天气、噪音、人造物等会影响到项目场地的因素以及在现场勘测时所拍的照片。

(Figure 3.9) A context analysis for the University of Arizona Poetry Center uses a base map overlayed with annotations regarding climate, noise, man made features and digital photography taken during onsite exploration.

我们还是推荐将多份基础地图拷贝夹在一个合适尺寸的硬质界面上，便于记录观察到的信息。这些地图一般可以从地方政府的相关机构或是道路管理部门获得。数码照片用来记录单独的元素以及动态的街道生活、交通状况等信息是非常有用的（图3.9）。

---

census data. Similar to site exploration, for the context analysis we recommend utilizing multiple copies of a base map of suitable scale mounted to a hard surface for recording observations. This map will generally be available through local governments through their utility or road departments. Digital photography is very useful for documenting individual elements as well as the dynamics of street life, transportation, etc (Figure 3.9).

场地
Place 83

　　一个利用场地分析来发展场地设计概念的很好的例子就是作为亚利桑那大学帕克学生中心利用率不高的重新研究的一部分，我们对场地文脉进行了研究。由此我们注意到以下几点：①建筑的标识系统不够充分，难以吸引学生；②周边的建筑给接近学生中心设置了障碍。正如我们前面所说的，场地记录同场地分析之间的区别就是：一个是简单地记录下已有的客观条件（图3.10），一个是提出自己的观点（图3.11）。

An example of using site analysis to develop site concepts occurred while studying the context as part of rethinking the under utilized Park Student Center at the University of Arizona in Tucson, Arizona. It was noted that ① signage was insufficient to attract students and that ② adjacent buildings formed a barrier discouraging access. As mentioned, the difference between site documentation and analysis is the difference between simply documenting existing conditions (Figure 3.10) and presenting opinions (Figure 3.11).

（图3.10）这张帕克学生活动中心文脉分析图利用合成的照片显示了周边设施是如何将潜在的客户同建筑物隔离开的。
(Figure 3.10) The Park Student Union context analysis shows how the facility was barricaded from potential customers through the use of simple diagramming combined with digital photography.

可能性
POSSIBILITIES
UNIVERSITY OF ARIZONA PARK STUDENT UNION
亚利桑那大学公园大道学生活动中心

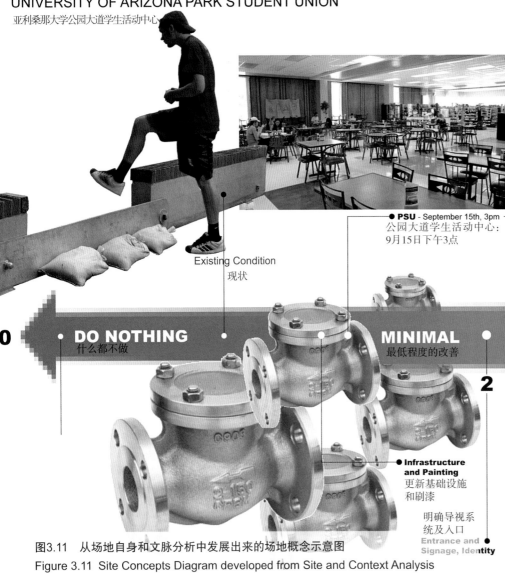

- Highland Market - September 15th, 3pm
  高地市场 9月15日下午3点
- Existing Condition 现状
- PSU - September 15th, 3pm
  公园大道学生活动中心：9月15日下午3点
- Coronado, Ariz new 5th Street Family and Co
- Infrastructure and Painting 更新基础设施和刷漆
- Entrance and Signage, Identity 明确导视系统及入口
- STUDE 学生交济

0 — DO NOTHING 什么都不做 — MINIMAL 最低程度的改善 — 2 — ATTRACT 吸引力

图3.11　从场地自身和文脉分析中发展出来的场地概念示意图
Figure 3.11　Site Concepts Diagram developed from Site and Context Analysis

设计任务示意图总结了亚利桑那大学帕克学生活动中心改造的各种可能性。这涵盖了各种程度的改造，从什么都不做、维修基础设施、通过多种途径吸引学生，直到最终将建筑转变成一个学校里的文化中心，如果能够得到足够多的资助，我们希望创造出一种基于多种语言同声传译系统的新建筑类型，我们称之为"翻译剧场"。

Program diagram summarizing possibilities for remodeling the University of Arizona Park Student Union Center. These ranged from doing nothing, to repairing infrastructure, attracting more students through various means, to transforming the building into a cultural centerpiece for the university and ultimately, given enough funding, inventing a new building type based upon simultaneously translating multiple languages in what we call the "translatorium".

学生活动中心周边的建筑
ra, Kaibab-Huachuca residence halls,
La Aldea graduate student housing,
ciences/McClelland Park

跨学科合作
INTERDISCIPLINARY COLLABORATION

Showcase 橱窗展示
agricultural 农业
research 研究
experiments 实验

8

The "**Translatorium**", as conceived by Line and Space, connects Students and Faculty to the International World
线和空间设想的"翻译大厅"将学生和教职员工同国际社会联系在一起

TRANSFORMATION
转换

REINVENTION
重新创造

M
OPP

4

NEW FACILITIES
新的设施

● Cuisines of the World
世界的美食

12

A CULTURAL CENTERPIECE FOR THE UNIVERSITY ●
校园中的一个文化中心

头脑风暴室
THINK TANK

帐篷
KAMP

● **CROSS POLLINATION OF IDEAS** 想法的传播
MAGNIFIED BY DIVERSITY 由多样化产生的放大效应

CHANGE

RIZONA DAILY WILDCAT
亚利桑那大学日报

CUISINE
烹饪

Photo Credits: L

LINE A
627 E. SPEED

地图

下面是几种会对你有帮助的地图类型：①各种比例尺度的谷歌地图，上面会显示可能会对场地有影响的交通系统、地面之上的市政设施，以及自然和人造物体。需要注意到的是各种接近场地的交通模式（自行车、小轿车、出租车、公交巴士、有轨电车、地铁和火车），各种到达及离开场地的节点，例如车站或终点站的位置、进入和离开的坡道、街道和高速路的容载力以及可能产生的噪音强度。②商业旅游地图（图3.12）(可能会被谷歌地球所取代)，能够提供一系列有用的信息，例如：主要的交通路线、可看到风景的次级路线、考古现场、历史遗迹或是休闲场地、国家公园、野生动物公园和保留地、行政区域间的边界、所有的道路、高速公路、铁轨、道路的距离、高程信息、地形、水文信息，大量的物理特征标识例如山脉、山谷和山峰，旅游信息以及大量游客会关注的兴趣点。更多的信息可以参看这个网址：http://www.maps.com。③类似的有市政府所印制的旅行地图，会提供大量的信息，一般还是免费的。④美国国家地理杂志地图一般关注的是环境、文化和人口统计方面的信息。⑤区域级别的政府所提供的地图经常可以在高速公路沿线免费获得，它对于了解周边建筑的相对经济价值会有些帮助，一般上面会印制区域或是不同产权间的边界并标示出地役权。⑥规划文件例如一个区域未来的发展规划图可以帮助我们预见这个区域的未来发展方向。

（图3.12）商业化的旅行地图在一个单独压缩的空间里提供了丰富的信息。
(Figure 3.12) Commercially available travel maps provide a wealth of information in a single compact package.

Maps

Types of maps that you will find useful: ① Google Earth maps at various scales show transportation that may affect your site, locate above ground utilities, as well as natural and man-made features. To be noted here are proximity of various transportation modes (bicycle lanes, auto/taxi, bus, trolley/subway and train), as well as access/egress in terms of location of stations, terminals, on and off ramps and street and highway capacity and intensity of noise generation. ② Commercial tourist maps (Figure 3.12) (probably being replaced by Google Earth), provide an array of useful information such as: Major transportation routes, Scenic secondary routes, Archaeological, Historical and Recreational sites & ruins, National Parks, Wildlife Parks & Reserves, International & Provincial borders, all roads, highways, trails and road distances, Elevation info, Hypsography and Hydrography, Extensive labeling of physical features such as mountain ranges, valleys and peaks, Tourist Information, and Points of Interest. See more at: *http://www.maps.com/* ③ Similarly, Municipal and government travel maps provide a wealth of information, generally at no cost. ④ National Geographic maps often focus on the environment, culture and demographics. ⑤ Municipal or Regional Government maps, often available on line include assessor maps useful in understanding relative economic value of nearby structures and showing property boundaries, areas and easements. ⑥ Planning documents such as plans for an area's future allow one to understand how growth is foreseen. ⑦ In the same way, municipal

⑦与此类似，市政府的功能分区图会指出场地周边的某个区域可能或者不会建造某种功能类型的建筑。⑧交通平面可以提供的信息包括现有的路线、各种不同体系的交通系统的容量和未来可能的增长。另外非常重要的是，它可以帮助我们判断出建造场地可能会产生的潜在问题。⑨市政设施地图非常重要，它可以帮助我们理解市政设施的管网线路、容量以及预测未来的增长，包括污水、供水、电力（主干线和次级干线的电力传输和分配）和通信。

最后，最好的低成本地图（和报告）资源之一就是美国和中国的国家政府（以及其他国家的相同机构）所提供的大量的信息，包括：⑩中国国家统计局或美国人口普查局等。⑪中国地质调查局、美国内政部、美国地质勘探局（USGS）。出版物包括来自USGS网站的地形图、降雨模式，科学出版物包括风险图（地震和火山），卫星拍摄到的航拍照片。⑫中国水利部和美国陆军工程兵团所提供的信息，包括水资源、洪泛区域、环境保护（与美国国家环境保护局一起）等（图3.13）。

当建筑师研究文脉的时候，最基本的就是要在一张文脉基准地图上找出地理条件、周边建筑物和它们所产生的影响以及基础设施等信息。请记住所有这些主题方面的信息都不是在场地中获得的。

(图3.13)美国地质勘探局针对一块特定的地质区域发布的报告编目的一部分如下：

- 简介
- 目的和范围
- 研究领域的描述
- 降雨模式和区域地表水
- 回收
- 研究内容描述
- 设计
- 致谢
- 湿地水文地质背景
- 区域水文地质信息
- 水文地质分类法
- 盆地地层构造
- 亚湿地地层构造
- 湿地流域地表水流动模式
- 湿地水文地质概况

(Figure 3.13) Partial Table of Contents of USGS publication focused on a single geographic area

- Introduction
- Purpose and Scope
- Description of Study Area
- Rainfall Patterns and Regional Ground-Water Withdrawals
- Description of Study
- Design
- Acknowledgments
- Wetland Hydrogeologic Setting
- Regional Hydrogeology
- Hydrogeologic Methods
- Basin Stratigraphy
- Sub-Wetland Stratigraphy
- Ground-Water Flow Patterns in Wetland Basins
- Overview of Wetland Hydrogeologic Settings

Zoning Maps point out possibilities for the type of buildings that may (or may not) be built nearby. ⑧ Transportation plans provide information on current routing and capacities of various types of transportation including future growth; invaluable for predicting potential interference with a given property. ⑨ Utility Maps are invaluable in understanding routing, capacities and predicted grown for sewer, water, power (Transmission and distribution of primary and secondary lines), and communications.

Finally, one of the best sources of low cost maps (and reports) is the US and Chinese Federal Governments (and other countries equivalents) vast resources and include ⑩ the National Bureau of Statistics of China / US Census bureau and ⑪ China Geologic Survey / US Department of interior, US Geologic Survey (USGS). Publications include (from USGS Website) Topographic Maps, Rainfall patterns, Scientific Publications including Hazard mapping (earthquakes and volcanoes), Aerials from Landsat 8. ⑫ China Ministry of Water Resources and United States Army Corps of Engineers provide information on Water resources, Flood Plains, Environmental protection (in conjunction with the US Environmental Protection Agency (EPA), etc (Figure 3.13).

When the architect is studying the context, at a minimum it is important to identify Geography, buildings and their influence, and infrastructure on a contextual base map. Remember these topics are all offsite.

信息的表达

场地照片可以排成一长条来展示（360°全景图），多张图片可以单独展示，附在地图上一起展示或是以视频的格式进行展示（图3.14）。场地文脉示意图可以单独展示（图3.10）或是结合场地基准地图一起展示（图3.9）。另外，一张复合型的场地示意图可以记录多种信息，包括所有的测绘数据、自然及人造因素（图3.7）。

Presentation of Information

Site photography can be presented in strips (360 degree panoramas), multiple images either alone or attached to maps or video format (Figure 3.14). Site context diagrams may be alone (Figure 3.10) or combined with the site base map (Figure 3.9). Additionally, composite site diagrams that document all survey data, natural forces and man made forces can be on a single map (Figure 3.7).

South

(图3.14)可以将多张图像拼合在一起形成360°的全景图作为设计师设计时的参考信息。
(Figure 3.14) 360 degree panoramas can be stitched together from multiple images and annotated to inform the designer.

场地的选择

作为建筑师，我们可能会被要求在场地的选择上给出建议，或是在一个很大的产权地中，或是在几个不同地点的场地中做出选择。

我们在这里讨论的所谓场地选择问题，是指在产权的归属已经确定的前提下，业主所拥有的可开发土地远大于建筑实际建造用地需要的规模（图3.15）。这种情况对于政府机构或是大的组织来说并不罕见，他们一般都持有很大面积的土地，例如公园或者是校园，规模往往远大于他们项目开发的实际需要。图森市附近的亚利桑那索诺拉沙漠博物馆项目和内华达州拉斯维加斯西边的红石峡谷国家保留地中的沙漠学习中心项目就都是这样的情况。我们选择来介绍这两个案例的原因是，在选择场地的时候，第一个案例为保护土地建立了一个有用的规则，而第二个案例则解释了为什么这个规则可能会被破坏掉。

这些程序以及它们相对不大的规模（在一个相对大一些的产权地中）一般来讲对于比较多个分散的场地是比较有用的方法。当然，这些方法必须要结合一些更广泛的规划问题一起来使用，例如场地的成本比较、场地的开发费用、税收方面的负担和收益（来源于一些政府的激励机制）、场地污染、环境方面的影响、筹募资金的机制，包括洪水和强风在内的气候条件、场地排水情况、人口密度、公众安全、距离公共服务设施的远近、市政设施（污水及雨水排放、电力管线、天然气管线、可饮用水供给、二次循环用水、垃圾处理以及通信）的容量及它们的成本、场地的进出入方式、可用的停车位、可用的公共交通以及许多其他的问题。

Site Selection

As architects, we may be asked to advise in site selection either within a specific large piece of property or among a number of different locals.

Site selection, as discussed here, focuses on property that is already owned but may be much bigger than required for the building (Figure 3.15). This situation is not uncommon among government entities or organizations with large land holdings, such as parks or campuses, far in excess of what is required for their project. This was the case at both the Arizona Sonora Desert Museum (ASDM), near Tucson, Arizona and the Desert Learning Center (DLC) within Red Rock Canyon National Conservation Area, just west of Las Vegas, Nevada. Actual site selection methodologies are examined here for both of these projects. The reason that we are illustrating two cases is that example one establishes a useful rule for preserving land when choosing a site and the second explains why this rule may be broken.

These procedures and their limited scope (within a single larger property) are useful in creating a general approach for comparing multiple dispersed locals. Of course, they must be used in conjunction with more broad planning issues such as comparative site cost, site development expenses, tax burdens and benefits (incentives), pollution and ground contamination, environmental impact, funding mechanisms, weather patterns including flooding and high winds, drainage, population density, public safety, proximity to municipal services, utility (storm sewers, power, natural gas, potable water, reclaimed water, waste and communication) capacities and their costs, access and egress locally, regionally and nationally, availability of parking, availability of public transportation, and many, many more.

场地
Place   91

VS.

一个本地的建造场地通常都没有条件做多场地比较的工作，因为场地的大小就基本上等同于建筑占地面积的大小了。一个很明显的例子就是在高密度的城市环境中有着最小退线要求的一块建造场地。
A Local Site does not allow for comparison of multiple sites because the site footprint is roughly equivalent to the building footprint. An example would be a dense, urban site with minimal setback requirements.

一块比较大的用地例如公园或是校园一般提供给建筑师可以建造的面积都远大于实际项目所需要的占地面积。相应的，他们就应该建立起评判标准，就几块分散的场地进行比较从而甄选出最合适的建造场地。
A large piece of land such as a park or campus presents the architect with buildable area far in excess of what is required for the building's footprint. In response, they must establish criteria to evaluate several Dispersed Local Site options to determine the best place to build.

图3.15 本地的单一场地与本地的几块分散场地
Figure 3.15 Local Site VS. Dispersed Local Sites

案例一 场地选择工具

亚利桑那沙漠博物馆世界闻名，其中收纳了许多当地的动物和植物。在一片被妥善保护的茂盛的景观植被中，沿着步行的路径布置了各种不同的展览。场地中的地形起伏极为不同，有的地方只是微微隆起，而有的地方则是非常陡峭。如何在博物馆将近8hm²的产权用地中，在众多可以使用的建造场地里挑选出一处最适合建造我们的项目——一个新的餐厅和画廊综合体的场地，是一个挑战。

在开始挑选场地之前，非常重要的一点是要知道项目的建造实际需要占用多大面积的土地。这个面积是在设计任务制定过程中的对话部分结束后确定的，在实际需要基础上计算出的建筑面积大约是2000m²。一个已经被证实为行之有效的方法是将最有潜质的待选地段按照不同的对比项评分累加，最高分的地段作为最终的建造用地。

我们根据一个成功项目所要包含的关键要素发展出一系列的评价标准，这些要素包括市政设施管网的分布、服务性车辆进入场地的难易程度、与现有场地人流的关系、景观条件、植被的破坏情况等。基于这样的思路，我们创造出了一个叫作"场地比较研究矩阵"的工具，在上面我们列出所有待选场地及它们的研究对比项（图3.16）。每一个对比项旁边的数字即为这块场地在这一

Site Selection Tools – Example One

The ASDM houses a world famous collection of animals and plants from its region. Various exhibits are along walking trails within a well-preserved, highly vegetated, natural landscape. The terrain is extremely varied from gently sloping to extremely steep. The challenge was selecting the best building location for a new Restaurant and Gallery Complex from among a number of compelling, available sites within the museum's 8 hectares (20 acre) property.

Before the site search could begin, it was important to know how much land was needed for construction. This was determined after the dialog portion of the programming process when calculations based upon needs indicated that the actual building area would be approximately 2,000 square meters (20,000 square feet). The idea, which proved effective, was that the most promising potential locations would be studied and graded numerically according to various points of comparison. The highest scoring location was chosen to build the project.

A set of criteria were developed identifying critical aspects of a successful project including such attributes as location of utilities, easy access for service vehicles, relationship to established visitor circulation, views, destruction of vegetation, etc. With these bases in mind a tool called the "*Site Alternatives Matrix*" was created listing potential sites and the factors under which they were studied (Figure 3.16). The numbers beside each factor indicate how well the site scored for each criterion (one being best; five being unacceptable).

对比项上的评分（1代表最好，5则代表不可接受，即最差）。使用这样的方式，我们就可以很容易地对不同的场地进行比较。所有的评估项都会被平等对待，不会出现某一项比起其他项的重要性大出很多的情况。

我们研究了7块场地。设计任务书中是这样描述的："最终靠近展示花园的一块场地被挑选为建造场地，正如你可以在比较研究矩阵中看到的(图3.16)，对于自然植被的破坏和对于正在进行项目的干扰都是最小，而对于适用性、景观、地形、位置的远近、可使用空间等方面的评估都很积极。"

除了"场地比较研究矩阵"这种工具之外，在已经开发过的土地上定位的时候，为了保护现有的植被或是尽量减少未来对于地形的修正，航拍照片会被用来进行场地研究。从空中，特别是对于一些脆弱的生态系统来说，一些诸如车辙、路径和可以被辨识出的空地会在很多年里都保持那样的状态（图3.17）。当然，在这种相对干旱的土地上，一些看起来很小的树也许会超过50年才可以生长成熟。比如巨型仙人掌，一种在当地占据统治地位的物种，可以活到超过150岁。

作为一条规则：永远优先使用已经被破坏过的场地作为你的建造场地。当这条规则被纳入为评分标准的一部分时，场地的破坏越严重，它的得分就会越高。这就会倾向于避免去选择那些未经开发破坏的处女地，因为它们在场地破坏方面的得分为0。

In this manner, comparisons between the areas could be easily carried out. No one criterion was deemed significantly more important than others so all were weighted equally.

Seven sites were examined. The program states, *"An area near the demonstration gardens was chosen and as you can see from the matrix* (Figure 3.16) *destruction of natural areas and disruption to ongoing operations were minimal while serviceability, views, terrain, proximities and availability of space were all positive aspects."*

In addition to the Site Alternatives Matrix, aerial photography was utilized with the intent of locating already disturbed land in order to save existing vegetation and minimize geologic disturbance. From the air, especially in fragile ecologies, vehicle tracks, paths, and cleared areas are readily discernible and will remain that way for many years (Figure 3.17) and, of course, in arid lands seemingly small trees may take fifty years or more to reach maturity. The Giant Saguaro cactus, a dominant and well loved feature here can live to be more than 150 years old.

As a rule: Use the land which has been most disrupted by prior use for your building site. When this criteria is used as part of the grading, the more disrupted a place has been the higher score it will receive. This tends to move the discussion away from pristine sites which receive "0" scores.

（图3.16）场地比较列表，将亚利桑那沙漠博物馆项目所候选的7块分散的场地分成若干选项加以评分比较。

(Figure 3.16) The site alternatives matrix applies a numerical rating to the seven distributed local sites at the ASDM.

影响场地的因素 Site Forces	亚利桑那沙漠博物馆建造场地选址 ASDM Site Alternatives	国王峡谷岛 King's Canyon Island	国王峡谷/停车场：低处的 King's Canyon/parking lot: lower	国王峡谷/停车场：高处的 King's Canyon/parking lot: upper	高地/田屋/老熊（图森市区）区 Overlook/ramada/old bear enclosure	蜂鸟：饲养场 Hummingbird: rear	蜂鸟：前方较低处 Hummingbird: lower front	展示花园 Demonstration garden
	景观：自然趣味 Views: natural drama	1	4	2	3	4	4	2
	景观：人看 Views: people watching	3	3	3	1	3	1	3
	景观：现存的障碍 Views: impedement of existing（内部/外部，来自于展览的）(internal/external, from exhibits)	2	1	1	3	2	3	1
	后勤维护：分离 Servicing: separation	4	3	3	2	2	2	1
	后勤维护：分散送达 Servicing: delivery distraction	4	3	3	2	2	2	1
	施工：破坏程度 Construction: disruption	1	1	1	3	3	4	1
	市政设施：可利用性 Utilities: availability	2	2	2	2	2	2	2
	噪音：对内/对外 Noise: in/out	2	3	3	4	4	4	2
	排水：自然和屋顶排水 Drainage: natural and roof water	1	2	2	2	2	2	2
	接近度：主要的动线 Proximity: major circulation（假设延长入口）(assume extended entry)	4	2	2	2	1	1	3
	接近度：礼品店 Proximity: gift shop（假设位置相同）(assume same location)	4	3	3	3	1	1	2
	接近度：入口 Proximity: entry（根据4月24日的会议）(per meeting 4/24)	3	2	2	2	2	2	3
	接近度：出口 Proximity: exit（根据4月24日的会议）(per meeting 4/24)	4	2	2	3	1	1	3
	接近度：夜间的动植物展示 Proximity: nocturnal（根据4月24日的会议）(per meeting 4/24)	4	4	4	4	1	1	1
	场地：项目信息 Site: ASDM message（不涉及拆除）(does not address destruction)	1	3	3	3	4	3	1
	场地：操作的破坏程度 Site: disruption to operations	3	2	1	3	1	3	1
	场地：地形的处理难度 Site: difficulty of terrain（与施工有关的）(relative to construction)	3	1	1	1	1	1	1
	场地：植被的破坏 Site: vegetative disruption	3	1	1	3	1	4	1
	场地：对自然区域的破坏 Site: destruction of natural area	5	1	1	3	1	4	1
	场地：可利用的空间 Site: availability of space	2	1	1	1	4	1	1
	场地：带有景观的朝阳方位 Site: solar orientation with view	1	4	3	1	1	1	1
	总得分 TOTAL SCORE	57	48	46	51	43	47	32

图例　Key
1=最高或者最好的排名　1=Highest or best ranking
2=非常好　2=Very good
3=可以接受　3=Acceptable
4=很差　4=Poor
5=不可接受　5=Unacceptable

潜在的备选场地 Potential Sites

场地
Place
95

（图3.17）航拍照片中清晰地展示出场地中已经被侵扰的区域。
(Figure 3.17) Aerial photograph clearly showing disturbed areas of the site.

"作为一个原则：尽量优先使用那些已经遭受了最严重的侵扰的场地作为项目的建设场地。"

"As a rule: Use the land which has been most disrupted by prior use for your building site."

案例二　场地选择工具

第二个例子为打破第一个例子中所陈述的规则提供了一个令人信服的论据。在这个案例中，任务要求是在内华达州拉斯维加斯旁边80000hm²的红石峡谷国家保留地中确定一块120hm²的场地作为一个沙漠学习中心综合体（DLC）的建造用地。

当我们得知项目的大致建造地点将选定在奥利弗牧场历史保护区内时，第一件要做的事就是召集包括美国土地管理局、附近的居民以及拉斯维加斯市政府代表在内的项目相关利益方进行充分的对话，讨论中很重要的一部分就是建立起项目的目标，理解项目的实际需求，这样我们就可以计算出项目大致需要的面积是多少。

在设计任务制定期间，沙漠学习中心的主要作用是被当作克拉克郡学区内5年级学生全身心的实地体验莫哈维沙漠的场所。我们的想法是，当这些学生被集中在莫哈维沙漠生活5天之后，他们会从周边的自然美景中深受启发，并掌握在这样干旱的土地上生存的技巧，进而影响到他们在未来可以积极参与到对这片同他们息息相关的自然环境的保护行动中去。这样的价值理念将被传递给他们的父母，但是最重要的是，这一代的儿童将会把这样一种保护环境的责任感带进他们的成年生活中。

Site Selection Tools – Example Two

The second example presents a compelling argument for breaking the rule mentioned in Example One. Here, the requirement was to determine the location for a Desert Learning Center Complex (DLC) within a 120 hectares (300 acre) parcel itself located in the 80,000 hectares (200,000 acre) Red Rock Canyon National Conservation Area just outside of Las Vegas, Nevada.

Knowing that our general location for the DLC was within the historic Oliver Ranch holding, the first order of business was establishing project goals, needs and wants as part of an extensive dialog with the United States Federal Bureau of Land Management, nearby neighbors, and the City of Las Vegas at large in order to be able to calculate the approximate building area the project would require.

During programming, the Desert Learning Center was conceived as a total immersion, Mojave Desert experience for 5th graders attending school in the Clark County School District. The idea was that after five days living in the Mojave Desert, the young students would become so inspired by the beauty of the land and the skills that they learned about how to live in an arid environment that they would become stakeholders in the welfare of their place and active in its protection. This knowledge and ethic would be conveyed to their parents, but most importantly, a generation of children would take this responsibility for stewardship of the land into adulthood.

牧场内3块6000m²左右的地块被认为最具潜质，我们分别对其作了勘测，并将它们分别命名为"牧场地块"（图3.18）、"已破坏地块"（图3.19），和"景观地块"（图3.20）。最基础的工作就是归纳总结出场地中不同的对比项，并将它们填入一个场地评估表格中（图3.21），我们使用"＋"号表示正评价，"－"号表示负评价，而"○"则代表这一项的评价为中性。正如你从评估表中所看到的，对比评估标准涵盖了从使用已经被破坏土地的能力、所受临近高速公路的影响、捕捉周边壮美景色的能力，一直到最终的未来扩展的难易度等诸多方面。这种分级评估的方法可以作为在第一个案例中介绍的数字打分评估方法的替换方案，这种方法适用于难以获得精确的评估价值的情况。

通常来讲，正如我们在第一个案例中所提到的，我们推荐使用已经被破坏过的土地作为新项目的建造场地，因为这可以有效降低对环境的影响。但是在这个案例中，经过慎重地考虑我们决定破一次例。因为我们基本设计目标是希望儿童可以全身心地去体验自然环境，这当然不能在已经被破坏过的场地中实现，最理想的建造场地应该是那块"景观地块"。

Three sites within the ranch each around 6,000 square meters (60,000 square feet) and appearing to have high potential, were examined and designated the "Ranch Site" (Figure 3.18), the "Disturbed Site" (Figure 3.19), and the "View Site" (Figure 3.20). Primary attributes for each site were discerned and summarized in a Site Evaluation Matrix (Figure 3.21), using the symbols "+" for positive aspects, "–" to indicate negative characteristics, and "0" if the attribute being studied was neutral. As you will see from the Matrix, criteria ranged from ability to use previously disturbed land, to impact from the nearby highway, to ability to capture stunning views and finally, ease of expansion. This hierarchal grading method, an alternative to the numerical values illustrated in Example One, is efficient for working in teams where a more precise value may be elusive.

Normally, as in Example One, we recommend using land which has already been disturbed for new construction, since it reduces environmental impact. Here, after careful consideration we made an exception. Since a primary goal, the desire for the children to be immersed in nature, could not be met on the "Disturbed Site", the best place for this project to be built was the "View Site".

备选的场地 Potential Sites

评估标准 Evaluation Criteria	农场 Ranch	被侵扰过的 Disturbed	景观 View
使用之前已经被侵扰过的土地 Uses previously disturbed land	+	+	-
对场地自然元素的破坏最小 Minimal disruption of natural attributes of site	-	+	-
使用现有的道路 Use existing roadways	+	+	o
利用真实的沙漠生活经验启发教育 True "desert experience" inspires education	-	-	+
融入莫扎维沙漠环境的感觉 Sense of integration into the Mojave Desert	-	-	+
Proximity to shade provided by existing tress	+	+	o
最大化历史建筑的价值 Maximize value of historic buildings	-	+	-
西伯利亚榆木的影响 Impact of Siberian Elms	-	o	o
159号高速路对于景观的影响 Impact on view from Highway 159	o	-	+
对于破坏景观的159号公路的遮挡效果 Shielded from distracting view of Highway 159	o	-	+
威尔逊崖的景观（全景）View of Wilson Cliffs (panoramic)	-	-	+
威尔逊峰的景观 View of Wilson Peak	-	+	+
蓝钻石山有趣的景观 Interesting view of Blue Diamond Hill	o	o	+
东南方向"小童山"的景观 View of "kid-sized" hill to the southeast	o	+	+
水池山脊上的岩石造型 View of cistern ridge rock formations	o	o	+
有吸引力的前景 Attractive foreground view	-	+	+
对于西侧建筑物的遮挡效果 Shielded from distracting views of buildings to west	+	+	+
距离上课地点的远近度 Proximity to teaching venues	+	+	+
通达其他场地地点的难易度 Ease of accessibility(ADA) to other site venues	o	o	+
对于来自于市政设施噪音的隔离 Noise isolation from WH&B Facility	+	+	-
对于来自于159号高速噪音的隔离 Noise isolation from Highway 159	-	+	o
相关的预算影响 Relative budget impact	+	o	-
高于100年一遇的洪水线（待确认）Outside of 100 year floodplain(to be verified)	+	+	-
将光伏发电结合进建筑 Integration of photovoltaics into buildings	-	+	+
与市政设施结合的难易程度 Ease of tying into utilties	+	+	-
同废物处理间的关系 Relationship to waste treatment	+	+	-
施工通达性 Construction access	+	+	-
未来扩展的难易度 Ease of expansion	+	+	-

图例 Legend
+ 正面评价 Positive Aspect
- 负面评价 Negative Aspect
o 中立 Neutral Aspect
　不可接受 Not Applicable

这份表格用来验证备选场地能否符合NEPA的要求及满足工程合理性
This matrix is subject to verification for conformance with NEPA requirements as well as engineering feasibility

场地评估表
红石国家沙漠学习中心
Site Evaluation Matrix
Red Rock National Desert Learning Center

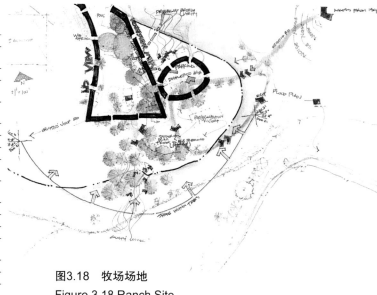

图3.18 牧场场地
Figure 3.18 Ranch Site

图3.19 已被干扰的场地
Figure 3.19 Disturbed Site

（图3.21）场地评估列表，将沙漠学习中心项目的三个分散的候选建造场地的各个方面用积极、消极和中立三个标准加以评价比较。
(Figure 3.21) The site Evaluation Matrix applies a positive, negative, or neutral rating to the three distributed local sites at the DLC.

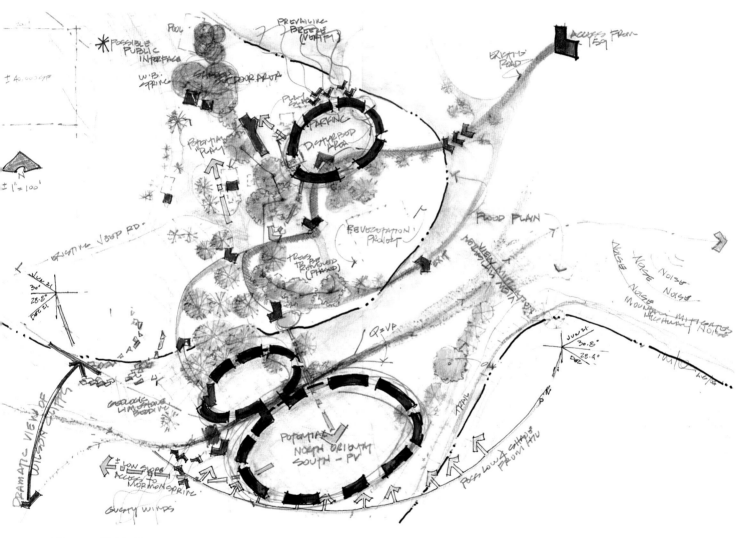

图3.20 景观场地

Figure 3.20 View Site

尽管景观场地没有利用场地中已经被侵扰过的部分,但这是可以接受的,因为设计任务书中要求儿童使用的设施应该完全融入自然环境当中,而这样的要求显然不能够在已经被侵扰过的场地中完成。

Although the View Site does not utilize the disturbed areas of the site for building, this was acceptable because the program required that the children that use the facility be completely immersed in a natural setting, which could not be accomplished on the disturbed site.

Site Documentation Checklist

Site Maps
- **Google earth**
  - Replaces aerials
    - Most disrupted land
  - Natural and man made features
- **GIS – topography**
- **Site survey**
  - Boundary with dimensions and bearings
  - Topography to appropriate accuracy
  - Vegetation
  - Natural features and anomalies
  - Drainage
  - Man made features
  - Utilities on site
  - Adjacent structures, roads and utilities
  - Easements
  - Encumbrances (tax liens, etc)

Site Forces
- **Natural**
  - Climate (sun, wind, rain, weather)
  - Hurricanes/tropical storms
  - Topography
  - Geology
  - Soil type
    - Core analysis
    - Stability/Angle of repose
    - Bearing value
  - Earthquakes
  - Volcanoes
  - Hydrology
    - Floods
    - Seasonal
    - 100/500 year
    - Drainage
  - Views (in/out)
  - Biology
    - Zoology
    - Botany

场地文案记录清单

场地地图
- 谷歌地球
  - 替换的航拍图
    - 被侵扰最严重的土地
  - 自然及人造物
- 地理信息系统
- 场地测绘
  - 带有尺寸及退线要求的场地边界信息
  - 达到恰当精确度的高程信息
  - 植被
  - 自然特征及反常现象
  - 排水
  - 人造物
  - 场地中的市政设施
  - 临近的构筑物、道路和设施
  - 地役权
  - 权利负担（税收优先权等）

场地影响因素
- 自然的
  - 气候（日照、风、雨、天气）
  - 龙卷风或热带风暴
  - 高程
  - 地质
  - 土壤类型
    - 地质采样岩心分析
    - 稳定性或休止角度
    - 承载力
  - 地震
  - 火山
  - 水文
    - 洪水
    - 季节性
    - 100年或500年
    - 排水
  - 景观视线（看进或看出）
  - 生物学
    - 动物学
    - 植物学

☐	Site vegetation	☐	场地植被	
☐	Types	☐	类型	
☐	Endemic/invasive	☐	原生的和入侵的	
☐	Patterns	☐	模式	
☐	Anomalies	☐	异常的	
☐	**Man made**	☐	人造的	
☐	Hydrology/drainage	☐	水文和排水	
☐	Storm drainage	☐	降雨排水	
☐	% slope	☐	找坡角度	
☐	Transportation	☐	交通	
☐	Airports	☐	机场	
☐	Noise	☐	噪音	
☐	from flight path	☐	经过航道	
☐	takeoffs and landings	☐	起飞和降落	
☐	Proximity/ease of access	☐	接近程度和通达的便利性	
☐	Trains	☐	火车	
☐	Noise	☐	噪音	
☐	Station access/egress	☐	出站和进站	
☐	Track	☐	轨道	
☐	Proximity/ease of access	☐	接近程度和通达的便利性	
☐	Buses, trolleys and subways	☐	公交、有轨电车和地铁	
☐	Noise	☐	噪音	
☐	Station access/egress	☐	出站和进站	
☐	Track	☐	轨道	
☐	Proximity/ease of access	☐	接近程度和通达的便利性	
☐	Future plans	☐	未来计划	
☐	Schedules	☐	日程表	
☐	Door to door transportation	☐	门到门的交通	
☐	Parking access/egress	☐	停车场地进出	
☐	Bicycles	☐	自行车	
☐	Limo's and Cabs	☐	专车和出租车	
☐	Private Auto	☐	私人轿车	
☐	Roads	☐	道路	
☐	Pollution	☐	污染	
☐	Noise	☐	噪音	
☐	Surrounding structures	☐	周边的结构物	
☐	Use	☐	使用	
☐	Shadows	☐	阴影	
☐	Wind	☐	风	
☐	Traffic	☐	交通	

- [ ] View obstruction
- [ ] Light pollution
- [ ] Site disturbances
- [ ] Utilities and capacities
- [ ] Communications
- [ ] Cable, telcom, wireless
- [ ] Sewer
- [ ] Water
- [ ] Recycled
- [ ] Utility wide
- [ ] On site - gray
- [ ] Potable
- [ ] Storm drainage
- [ ] Harvested
- [ ] Power
- [ ] Access/egress
- [ ] **Ecological implications**
- [ ] **Regulations**
- [ ] Zoning ordinance constraints
- [ ] CCRs
- [ ] **Goals**

Context

- [ ] **Maps**
- [ ] Google earth – various scales
- [ ] Google maps
- [ ] Travel maps
- [ ] Privately published
- [ ] National Geographic
- [ ] Municipal and Government Tourist maps
- [ ] Federal Government (US examples)
- [ ] USGS
- [ ] Landsat (worldwide)
- [ ] topographic and boundary maps at various scales
- [ ] Federal Lands
- [ ] Corps of engineers
- [ ] Water flow
- [ ] Flood planes
- [ ] GIS – topography

- [ ] 遮挡视线的因素
- [ ] 光污染
- [ ] 场地干扰因素
- [ ] 市政设施及其容量
- [ ] 通信设施
- [ ] 电缆、无线电信信号
- [ ] 下水道
- [ ] 水
- [ ] 循环使用的
- [ ] 使用范围
- [ ] 场地中的灰水
- [ ] 可饮用水
- [ ] 降雨排水
- [ ] 雨水收集
- [ ] 电力
- [ ] 进入和离开
- [ ] 生态影响
- [ ] 规定
- [ ] 功能分区条例的限制
- [ ] 立约
- [ ] 目标

文脉

- [ ] 地图
- [ ] 谷歌地球——不同比例尺度的
- [ ] 谷歌地图
- [ ] 旅游地图
- [ ] 私人发行的
- [ ] 国家地理
- [ ] 市政府旅游地图
- [ ] 联邦政府（例如美国）
- [ ] 美国地质调查局
- [ ] 陆地资源（全球范围内）
- [ ] 地形与边界不同比例尺度的地图
- [ ] 联邦土地
- [ ] 工程师团队
- [ ] 水流
- [ ] 洪水冲击面
- [ ] 地理信息系统

☐	Private or Corporate Utility Maps	☐	私人或机构的市政设施地图
☐	Power, Communication	☐	电力、通信
☐	Growth	☐	增长
☐	Municipal or Regional Government maps	☐	市政府旅游地图
☐	Assessor maps	☐	评估用地图
☐	Planning	☐	计划
☐	Adopted growth plans	☐	被采用的增长计划
☐	Plans for the future	☐	未来的计划
☐	Zoning	☐	功能分区
☐	Streets and highways	☐	街道和高速公路
☐	Public transportation routing	☐	公共交通道路
☐	Utility maps: Sewer, Water	☐	市政设施地图、下水道、上水
	Surroundings		周边环境
☐	Transportation impact	☐	交通影响
☐	Airports	☐	机场
☐	Noise from flightpath	☐	从航道传来的噪音
☐	Ease of access	☐	场地通达的便利性
☐	Direct transportation	☐	直接交通
☐	Trains	☐	火车
☐	Buses, trolleys and subways	☐	公交大巴、有轨电车和地铁
☐	Schedules	☐	日程表
☐	Future plans	☐	未来的计划
☐	Building impact	☐	建筑的影响
☐	Scale	☐	尺度
☐	Shadow impact	☐	阴影的影响
☐	Potential hazard impact	☐	潜在的灾害所会带来的影响
☐	Earthquakes	☐	地震
☐	Fire	☐	火灾
☐	Utilities	☐	市政设施
☐	Location	☐	位置
☐	Capacities	☐	容量
☐	Public building	☐	公共建筑
☐	Hospitals	☐	医院
☐	Emergency rooms	☐	急诊室
☐	Helipad	☐	直升机停机坪
☐	Ambulances	☐	救护车
☐	Schools	☐	学校
☐	Museums	☐	博物馆
☐	Religious	☐	宗教场所
☐	Restaurants	☐	餐馆

☐	Financial	☐	金融
☐	Industry	☐	工业
☐	Heavy	☐	重工业
☐	Light	☐	轻工业
☐	Warehousing	☐	仓储
☐	Shopping	☐	购物
☐	Retail	☐	专卖店
☐	Grocery	☐	杂货店
☐	Traffic generation	☐	交通生成
☐	Parking	☐	停车
☐	Neighbors/neighborhoods	☐	邻居或临近社区
☐	Demographics	☐	人员构成
☐	Age	☐	年龄
☐	Race	☐	种族
☐	Religion	☐	宗教
☐	Income	☐	收入
☐	Education	☐	教育程度
☐	Culture	☐	文化
☐	History	☐	历史
☐	Parks	☐	公园
☐	Public Safety	☐	公共安全
☐	Fire	☐	消防
☐	Police	☐	警力
☐	Public Health	☐	公共健康
☐	Power generation	☐	电力生成
☐	Sewage treatment	☐	污水处理
☐	Travel abroad	☐	出国旅游
☐	Cultural	☐	文化的
☐	Political issues	☐	政治问题
☐	Disease	☐	疾病
☐	Mosquito borne	☐	蚊媒
☐	Water	☐	水
☐	3d Models	☐	三维模型
☐	Community planning tools	☐	社区规划工具
☐	Mass models	☐	体量模型
☐	Sketch up	☐	草图大师模型
☐	Utilities and services	☐	市政设施及服务

Presentation of information
- **Site Photography**
- **Site Diagrams**
    - Composite
        - Mike's
        - Poetry Center
    - View diagrams
        - Old line and space circles
        - Photographic
    - Context diagram

Site Exploration
- **Materials and equipment needed**
    - Foam core boards
    - Clips
    - Map blanks
    - Topo map
    - Context map
    - Aerial map
    - Appropriate clothing – hats, footwear, raingear
    - Water
    - Markers, Pens and Pencils (blues, red, green and black)
    - 1 kg Sledge Hammer
    - Wood stakes
    - Wood lathe
    - Surveyor's ribbons
    - Compass
    - GPS
    - Communications – cell or hand held
    - Digital cameras/batteries
    - Plant and animal field guides
    - Insect repellent

场地 Place 105

信息表述
- 场地照片
- 场地示意图
    - 综合信息
        - 麦克的原始记录
        - 诗歌中心
    - 景观视线分析示意图
        - 老的空间关系示意图
        - 照片所记录的信息
    - 文脉示意图

场地踏勘
- 需要的材料和设备
    - 试波板（KT板）
    - 夹子
    - 空白的地图
    - 地形图
    - 文脉地图
    - 航拍图
    - 合适的衣服、帽子、鞋类、雨具
    - 水
    - 马克笔、墨水笔和铅笔（蓝色、红色、绿色和黑色）
    - 1kg重的锤子
    - 小木桩
    - 木条
    - 勘探用布带
    - 指南针
    - 卫星定位仪
    - 通信设备——手机或对讲机
    - 数码相机及电池
    - 熟悉当地植物和动物的向导
    - 驱蚊水

对话
Dialog

对话是设计师主导的设计任务制定（或是设计任务确认）工作的核心内容。在这本书中，对话是指项目利益相关者同建筑师之间相互讨论功能需求并且交换看法以期确定出建筑及环境的设计目标的过程。

Dialog is the companion to **place** as the heart of the designer directed programming (or program confirmation) effort. Throughout this book, dialog refers to stakeholders discussing needs and wants with the architect as well as exchange of ideas leading to formulation of environmental and building goals.

第一部分　准备

这个章节的第一部分，解释了开始对话前所需要做的准备工作，包括准备会议地点、安排主持人和发言人、确定会议或研讨会主题、撰写会议议程以及邀请项目利益相关者参会。尽管这些方面的问题看起来都似乎放在前面关于开始的部分讨论更合适，但它们和对话之间纠缠不清的关系使得放在本章讨论也很合理。

正如前面的开始一章的最后部分所提到的，结合着创建设计任务制定活动的清单和准备各种相关会议的日程表，建筑师将会任命一名项目经理，他负责统领设计任务制定的工作。这位项目经理将会提前同项目决策人讨论以确保会议地点都可以正常使用。

会议

会议可以有各种规模和类型，可以是10~100人较大群体间以座谈会或是研讨会的形式互动，也可以是2~10人较小规模的沟通，还可以根据不同个体分享信息的习惯进行一对一的交流（图4.1）。通常情况下我们知道，诸如董事会成员、总监、部门高管等一些领导位置上的人会参加集体讨论，但他们更愿意直接将信息传达给建筑师。

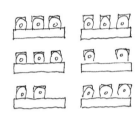

一对一
One-on-One

小规模聚集空间
研讨会格局
Small Gatherings
Seminar Format

大型团体
课堂格局
Large Groups
Classroom Format

（图4.1）准备对话非常重要的第一步就是建立起研讨会的框架日程。
(Figure 4.1) An important first step in preparing for dialog is establishing the framework for meetings and workshops.

Part I – Preparation

The first part of this chapter, explains getting ready to begin dialog including procuring meeting venues (locale of event), arranging facilitators and lecturers, setting meeting/workshop topics, writing session agendas and inviting stakeholders. While these subjects might well fit in the discussion regarding **beginnings**, their entwined connection to dialog makes their location in this section logical.

In conjunction with creating the complete list of programming activities and preparing the schedule for implementing these activities as explained at the end of the chapter **beginning**, the Architect will appoint a *Project Manager* who will lead the programming effort. This Manager will begin making arrangements to secure the meeting venues discussed previously with the Decision Maker.

Meetings

Meetings can include large groups (10 – 100 persons) interacting in a workshop or seminar format, small gatherings (2 - 10 persons) and one-on-one sessions accommodating various individual's propensity for sharing information (Figure 4.1). As a general rule we know that that board members, directors, department heads and persons in similar positions will want to convey information directly to the architects as well as participating in larger sessions.

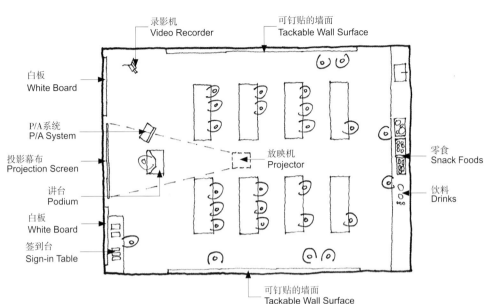

（图4.2）这张平面示意图展示的是一个典型的75m²空间内的设计任务讨论会的布置方式。
(Figure 4.2) Plan diagram showing typical layout of a seminar style programming session in a 75 square meter (800 square foot) space.

每个参加会议的人大约需要1.5~2m²的空间，另外我们还要留出足够的地方给会议的主持人、大张写字板、绘图板等。

地点

我们将下述地点定义为传统意义上正规的会议地点，它们包括会议室、董事会会议室、酒店中的会议室、会议中心、大学中的教室、机构、大学或是市政礼堂。这些地点用来决定一些项目的基本事实会比较有效，这其实就是回答这样一些问题：谁？什么？为什么？并且，在哪儿？还有，多少？进一步的，多少钱以及多大？

Meetings and workshops are generally organized in a space that allows for 15 - 20 square feet (1.5 - 2 square meters) per participant plus enough room for a moderator, large format note takers, graphic panels and notes.

Venues

We define formal meeting places as those traditional meeting spaces such as conference rooms, board rooms, hotel meeting rooms and conference centers, university classrooms, and institutional, university or civic auditoriums. These spaces work well for determining building facts; that is, the answer to the questions: Who? What? Why? And, where? As well as, How many? And, how much and how large?

在会议安排中包含有通信技术方面的要求,例如音频或视频设备(大尺寸的投影幕布、数字投影仪、带有方便观众参与的独立麦克风的公共广播系统)、影像及声音录制设备、座椅的数量(标准尺寸是50cm宽的可堆叠座椅)和类型、讲台及台阶、足够多的可以钉或是粘贴的墙面用以展示信息(对于较大的会议,最小的尺寸是水平向距离为12~24m,竖向高度是1.2m高,并且应该安置在比较舒适的观看高度上),大张的白板(2张2.4m×1.2m),迎宾或签到台和需要的食物及软饮(图4.2)。考虑到零食和饮料(糕点、士力架、水果、冰、水、碳酸饮料、咖啡和茶)需要在早上及下午持续供应并且不断补充,我们可能还需要单独的台面或是桌子。饮料需要同干的物品分开摆放。冷饮需要存放在冰柜中。咖啡和茶当然是需要有单独的容器。一次性杯子、玻璃杯、盘子和配有餐巾纸的餐具应该摆放整齐。对于全天的活动,一般在设计任务制定的过程中,一日三餐要妥善安排。三明治和沙拉应该就可以满足要求。这样做并不仅仅是回馈给参与者一顿午餐那么简单,其实它也是一种保证大家在下午的活动环节还可以聚拢在一起的方法。

一些不正式的会议场所,指的是那些可以更加休闲放松聚会的场所,例如利益相关者的办公室、咖啡店、咖啡馆、图书馆里的社区活动室、临近学校里的教室或是礼堂,这些地方都会使参与者感觉更加舒适,也就更愿意交谈。这类场所特别适合用来讨论理解场地及文脉,草拟设计目标以及分享意见。一些硬件方面的要求同上文中描述的类似。

Included in this arrangement will be technological requirements for communications such as audio/visual equipment (large projection screen, digital projector, PA system with freestanding microphone stand for audience participation), video and audio recording machines , seating both numbers(standard .5m (20") wide stackable chairs) and type, podium/lectern and risers, sufficient tackable or tapeable wall surface for display of information (for large meetings a minimum of 12m – 24m horizontally (40-80 linear feet), a minimum of 1.2m (4') high, and at a comfortable viewing height), large whiteboards (2ea – 2.4m x1.2m (8' x 4'), greeting/sign-in tables and refreshment and food needs (Figure 4.2). Consider that snacks and drinks (pastries, health bars, fruit, ice, water, soda, coffee and tea) should be continuously available and replenished in the morning and afternoon and will need a separate counter or table. Liquids should be well separated from dry items. Cold drinks will be kept in ice chests. Coffee and Tea of course need appropriate dispensing vessels. Disposable cups, glasses, plates, and utensils along with paper napkins should be stocked. For all day sessions, the norm during programming, catering must be arranged and included daily. Sandwiches and salads will suffice. Not only does providing lunch reward participants, but it also keeps the group together for afternoon sessions.

Informal spaces, that is, gathering areas that are more laid back and relaxed such as a stakeholder's office, coffee shop, cafe, library community room, neighborhood school classroom or auditorium tend to make participants feel more comfortable and willing to talk. These kind of spaces work particularly well for understanding site and context, crafting goals and sharing opinions. Logistical requirements are similar to those mentioned above.

在我们的经验中，没有比在实际的建造场地中搭建一个帐篷为会议场所进行会议更能调动参与者的参与感和积极性的了（图4.3）。当然，好天气是会议成功的关键，所以一般来讲，这样的安排最好是在春秋季节，当然明智的选择是再准备一个备用地点以防万一。尽管不同的天气状况会不期而遇，但是所有的参与者都可以亲身体验场地中的各种影响因素，例如景观、植被、高程变化、气候以及噪音。

（图4.3）搭建在场地中的帐篷给项目相关利益者提供了一个直接接触场地的机会。

(Figure 4.3) A tent set up on the project site offers stakeholders an intimate connection with the site.

In our experience, there is no better venue, for creating excitement and participation of stakeholders, than having meetings in a tent set up on the project site (Figure 4.3). Good weather is essential for success so normally this option will work best in fall or spring, it is wise to have a backup space just in case. While weather can play an unwelcome role in this locale, all parties become intimately acquainted with site forces such as views, vegetation, topography, climate and noise.

对于其中的一个项目，线和空间事务所将原本两周长的设计任务制定环节一分为二，一周时间在学校礼堂进行较为正式的会议（图4.4），另外一周则在场地搭帐篷宿营（图4.5）。这两个不同地点的与会者分享意见及参与热情的差异就好像夜晚与白天。一个原本的异见者在积极参与了一次我们在现场宿营组织的研讨会之后很快变成了一名支持者。

为了帐篷会议可以成功举行，我们需要一台发电机。它所产生的电力可以为夜间的会议提供照明，并且可以带动风扇改善帐篷里的小气候。另一件我们

（图4.4）一些正式的场地例如这个学校的讲堂其实不难获得，并且可以提供必要的座椅和灯光。

(Figure 4.4) Formal spaces such as this school auditorium are easy to aquire and provide seating and lighting.

For one project, Line and Space split a two-week programming effort into one week in a University Auditorium (formal) (Figure 4.4) and one week in a tent on site. (Figure 4.5) The difference in enthusiasm and sharing between the two venues was like night and day. A major dissenter became very involved once we started meeting on site and soon became a supporter.

For the tent to be successful, an electric generator is required. This will provide power for lighting during evening meetings as well as the tempering of the microclimate with fans. One other thing to keep in mind is that tack surfaces for pinning up notes, maps, photos

需要记得的事是我们还需要搭建一个可以钉或是粘贴的墙面用以展示信息，例如便条、地图、照片或是意见卡。将100mm×100mm的木柱像围栏柱桩那样打入地面就可以轻松实现这样的功能。这样的木桩每隔2.4m安放一个，在顶部和底部分别用50mm×100mm连续的横向构件连接，上面再固定1.2m×2.4m×13mm的白板，这样就做成了一个十分经济有效的可以钉或是粘贴的展示板。使用普通的螺丝钉就可以轻松组装和拆卸这样的装备。在任何情况下，会议场地通过公共交通到达的便利性以及场地周围大量的免费安全的停车位都会帮助会议取得成功。

---

and comment cards must be constructed. This is easily accomplished with 100mm x 100mm (4" x 4") wood columns set into the ground like fence posts. These are spaced 2.4m oc with 50mm x 100mm (2"x 4") continuous horizontal members top and bottom to support 1.2m x 2.4m x 13mm (4'x 8' x ½") homosote board, an economical and excellent tack able surface. Screws should be used for ease of assembly and demounting. In all cases, a well-located venue that is easily accessible by public transportation and plentiful, secure, free parking helps insure success.

（图4.5）搭建在场地中的帐篷给项目相关利益者提供了一个直接接触场地的机会。
(Figure 4.5) A tent set up on the project site offers stakeholders an intimate connection with the site.

组织对话

一旦确定了会议的地点,项目经理的工作就会进入到下面的环节:①设定会议或研讨会的主题;②编写会议议程;③安排主持人及发言人;④邀请项目利益相关者参加会议。

在线和空间事务所,我们会为未来的设计任务文件制作一个目录表(图4.6),我们利用这样一种方式来确定研讨会的主题以及设置会议议程。为了能够理解目录表包含着什么样的内容,下面的段落纲要性地列出了当我们为设计任务文件收集信息时需要涵盖的主题及其排列顺序。

与我们的直觉感受相反,头两个章节——简介和设计方案,其实是最后完成的部分。为了有序组织研讨会,我们会先开始针对场地和文脉的讨论,接下来详细探讨关于人的问题,包括参观者、办公管理人员,进而引申至对于技术设备的特别要求,包括采用什么样的系统、如何移动以及怎样操作运行。根据项目的不同,设计任务中将会包含有一个章节专门来定义建筑的类型。举例来说,对于一个博物馆项目,愿景描述这样一个章节会勾画出设计的哲学,描述出设计的目标、主题和未来的发展方向。对于所有的建筑类型而言,一个房间和空间表格将会阐释下面这些问题:都有哪些空间?这些空间怎样组团分区?对于每一个空间来说最重要的功能是什么?哪些关键的问题之间存在着关联?在这一章,我们列出了一个典型的空间分析表格,其中标出了所有在这个空间中所需要的设备和家具,附带的基础平面可以帮助我们准确地判定空间的面积及尺寸以及由此形成的内部交通

Organizing Dialog

As soon as decisions regarding type and procurement of venues are completed the Project Manager will proceed with ① Setting meeting/workshop topics, ② Writing session agendas, ③ Arranging facilitators and lecturers, and, ④ Inviting stakeholders.

At Line and Space, a table of contents (Figure 4.6) is created for the future program document as a way of defining workshop topics and setting meeting agendas. In order to be able to understand what subjects are to be included in the table of contents, the following paragraph outlines the major topics and the order that we address them when gathering information that forms the basis of our programs.

Counter intuitively, the first two chapters, the **introduction** and **design approach**, are the last chapters written. For purposes of organizing the workshops, we begin with discussions regarding **site** and context followed by a detailed discussion of **people**, including visitors, administrators and staff, lead to the very specific requirements of the **facility** including systems, movement and operations. Depending upon the project, the program will include chapters specific to the building type. For example, in the case of a museum, **the interpretive vision** outlining philosophy, interpretive goals, themes and voice may go next. For all building types, a **rooms and spaces** chapter will determine what the spaces are, how these spaces may be grouped or zoned, what is important functionally for each space, and what critical issues are associated with each? In this chapter, we typically have a worksheet for each space identifying all equipment and furnishings needed,

0.0 Introduction
   0.1 Mission
   0.2 The AHS and the Southern AZ Division
   0.3 Role within the Community
   0.4 Executive Summary

1.0 Design approach
   1.1 Reinvention
   1.2 Goals
   1.3 What is the AHS?
   1.4 Codes and Regulatory Requirements
   1.5 Review and Approvals
   1.6 Unresolved Issues

Workshop Sessions Start Here

2.0 Site
   2.1 General
   2.2 Site Selection Criteria
   2.3 Site Diagram

3.0 People
   3.1 General
   3.2 Goals and Objectives
   3.3 Primary Audience (visitors)
   3.4 Visitor Type/Characteristics
   3.5 Administration. Staff and Volunteers
   3.6 Organizational Chart

4.0 Facility
   4.1 Introduction
   4.2 General Goals and Objectives
   4.3 Sustainability Goals and Objectives
   4.4 Movement
   4.5 Parking
   4.6 Operating Schedule
   4.7 Systems

5.0 Interpretive Vision
   5.1 Interpretive Philosophy
   5.2 Interpretive Mission and Goals
   5.3 Audience and Constituency
   5.4 Themes and Messages
   5.5 Voice
   5.6 Interpretive Approach

6.0 Rooms and Spaces
   6.1 Space Relationships
   6.2 Room/space sheets
   6.3 Area Summary Worksheet
   6.4 Relative Area Diagrams
   6.5 Interdepartmental Sharing

7.0 Schedule

8.0 Budget

9.0 Appendix

0.0 简介
   0.1 任务
   0.2 AHS组织和南亚利桑那分部
   0.3 在社区内担任的角色
   0.4 执行概要

1.0 设计方案
   1.1 重塑
   1.2 目标
   1.3 什么是AHS?
   1.4 相应的规范要求
   1.5 审核及确认
   1.6 没解决的问题

研讨会会议从这里开始

2.0 场地
   2.1 一般性的问题
   2.2 场地选择标准
   2.3 场地示意图

3.0 人
   3.1 一般性问题
   3.2 目标和目的
   3.3 基础的观众（访客）
   3.4 访客的类型及特点
   3.5 管理人员。正式职员和志愿者
   3.6 组织架构表

4.0 设施
   4.1 简介
   4.2 一般性的目标和目的
   4.3 可持续发展方面的目标和目的
   4.4 移动
   4.5 停车
   4.6 操作日程表
   4.7 系统

5.0 设计表述的视觉呈现
   5.1 设计表述采用的理念
   5.2 设计表述的任务和目标
   5.3 设计表述的听众和赞助人
   5.4 采用的方法及传达的信息
   5.5 声音
   5.6 设计表述的方案

6.0 房间及空间
   6.1 空间关系
   6.2 房间或空间列表单
   6.3 面积统计表
   6.4 相关的面积示意图
   6.5 空间交叉及分享

7.0 日程表

8.0 预算

9.0 附件

归纳总结的部分在对话结束之后书写完成

Summary Sections written after completion of dialog

目录部分应该在对话开始之前完成以帮助指导组织研讨会的日程安排

Table of Contects Sections written before dialog to organize workshop agendas

（图4.6）创建一个目录表来指导组织定义研讨会的日程安排，能够帮助建筑师更好地计划接下来同项目利益相关者们的对话。

(Figure 4.6) Creating a Table of Contents to organize and define workshop agendas allows the architect to develop ideas and structure the forthcoming dialog between stakeholders.

Session Five: Spaces
4:45pm – 5:30pm
confirmation of spaces/rooms. a detailed list
attached or adjacent exterior space. this li
attributes of each space such as function, co

（图4.7）上面是从实际的日程表中摘出的一个典型的议程描述
(Figure 4.7) Above is a typical session description from actual agenda.

流线（参见"文案"章节）。也可以很好地利用这一部分的工作来记录不同空间之间的关系，并用来检验相对的面积指标是否合理（参见"文案"章节）。

一旦我们完成了目录表，详细的讨论议程就可以被制定出来（图4.7）。议程非常详细，包含具体的讨论主题、会议时长以及会议开始时间。设计任务研

---

along with a preliminary layout, which allows the size of that area to be accurately determined when internal circulation is factored in (See document chapter). This is a good place to also document adjacencies among spaces and to examine relative area diagrams (See document chapter).

Once the table of contents has been completed, detailed session agendas are created (Figure 4.7). These agendas are very specific as to topics, duration, and time of day.

（图4.8）在同项目利益相关者一起讨论过后列出的所需空间。
(Figure 4.8) Required space list after review with stakeholders.

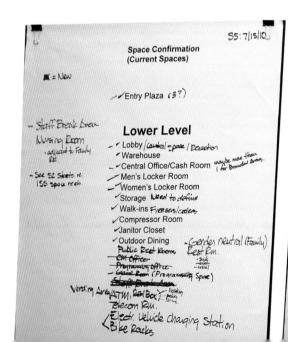
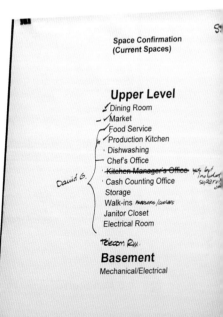

developed for all building spaces for the facility including
based upon need and want. priorities are established.
uration, etc, are not discussed here.

讨会通常会被设计为2~5天的讨论。一个复杂的项目可能设计任务讨论的时间很容易地就花费掉一个月（图4.9）。每个主题一般会讨论1~2个小时。一天的讨论一般涵盖4~5个部分，上午2个，下午2~3个。再加上晚上的一个会议就可以占满一天。一旦日程确定下来，就应该尽量严格执行，因为参加会议的人时间有限，他们只会去参加比较感兴趣或者是有话可说的讨论。

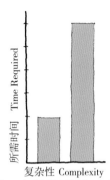

（图4.9）阐释项目复杂性与设计任务制定所需设计时间之间关系的示意图
(Figure 4.9) Diagram illustrating project complexity VS time required for programing.

Programming workshops will usually be designed to take place over the course of 2 to 5 days. Complex projects can easily take up to a month to program (Figure 4.9). Topics are generally discussed in 1 to 2 hour sessions. A day consists of 4 or 5 sessions; two in the morning and two or three in the afternoon. An evening meeting can round out the day. Once finalized, it is important that agendas are closely adhered to, as participants may have limited time and will attend only those sessions that they are most interested in and can contribute the best information to. For example, a discussion focused upon understanding staff needs, how they function and interact will obviously depend upon staff participation, while a site session will be enhanced by social scientists, geologists, biologists, hydrologists and others strongly related to the science of the place.

（图4.10）建筑师同项目利益相关者一起填写空间信息表。
(Figure 4.10) Architect filling out space information worksheets with stakeholders.

举例来讲，一个讨论员工需要的会议，要想理解这些功能空间怎样工作与互动，就明显依赖于员工的参与。而一个关于场地的讨论需要的则是例如社会学家、地质学家、生物学家、水文学家这些同场地科学紧密相关的人士。

一般来说，设计任务研讨会的第一天都会以一个项目的简单介绍开始，在随后的连续几天中，每一天都会简单概括一下前一天讨论的内容。为与会者提供1小时的午餐以及上下午各15分钟的休息时间是非常重要的。为了尽量让公众参与（甲方的决定），可以安排晚上的会议来回顾白天观察到的以及讨论过的特殊的主题。

一旦完成了会议议程的制定，我们就建议立即将议程送交给项目决策人来审阅。一经核准，马上联系相关人员以保证他们可以参加相应的讨论会。

对于每一次讨论会来说，有一个可以主持大家讨论并熟悉会议主题的人是很重要的，他可以提取相关的信息，控制群组的讨论不跑题。这个主持人可以是建筑师、旗下的员工、甲方的工作人员或者管理人员，甚至可以是外请的专家。关键的一点是为每一次会议找到合适的主持人并确保他们的档期与会议的时间匹配。

当邀请一些关键的项目利益相关者参加讨论时，非常重要的一点是一定要亲自联系他们以表诚意，这样可以增加他们参加会议的可能性。通常来说，这项工作的最合适人选是项目决策人。作为对此的支持，建筑师应该给每个与会者发送标有明确会议日程的请柬。在公开邀请的情况下，可以考虑采用报纸广告、张贴海报（在图书馆里）或者是使用其他形式的社会媒体。

Normally, the first day of a programming workshop begins with an introduction, and on successive days an overview of the prior day. It is important that a 1-hour lunch break is provided along with 15-minute breaks in the morning and afternoon. In order to involve the public (a client decision), evening sessions can be held recapping the day's observations and soliciting input on particular topics.

As soon as the agenda has been completed it is advisable to review it with the decision maker. As soon as it is approved contact participants in order to assure that they will attend the proper sessions.

A person familiar with the programming topic and who can lead group discussions, extract relevant facts and keep a group on the subject at hand is important to lead each session. This facilitator may be the architect, members of his staff, client staff members or administrators or outside experts. The key here is to recognize the appropriate person to assure they are committed in a timely way.

When inviting key stakeholders it is important to personally contact them to increase the probability of attendance. Generally, the best person to do this is the decision maker. In support of this the architect should send out personal invitations with the schedule clearly delineated and in the case of public invitations consider newspaper advertisements, posters (in libraries) and the use of various social media.

第二部分 互动

这个章节的第二部分解释了与会者之间如何互动,包括对于会议主题的讨论。特别的是,这一部分介绍了如何去理解项目的目标和需求。

一个设计任务研讨会一般开始都会由项目决策人(业主方)进行欢迎致辞,然后就是简要介绍一下讨论会的目的及第一天的议程安排。设计任务制定团队的成员应该清楚地知道自己的职责。因此找一些机会让会议参与者们在会场中走动一下,介绍一下自己及他们在项目中所扮演的角色,这会是很有意义的事情。每一天在上述部分结束后,新加入会议的成员会被介绍给大家认识,随后会回顾上一天讨论的内容,并简要介绍当天需要讨论的主题。最好是在开始第一个议题之前让与会者有充足的时间互致问候,修整落座。

作为背景介绍的一部分,我们会在会场到达区比较明显的地方放置一些尺寸相对较大的图片(有关文化、场地、艺术、使用者等方面)、地图(从大尺度到小尺度)以及大张的设计任务陈述展板,这些资料也可以持续地在之后的讨论中被作为参考(图4.11)。

PART II – Interaction

The second part of this chapter explains interaction among participants including topics for discussion. In particular, this section addresses understanding project goals, needs and wants.

A programming workshop will begin with welcoming remarks by the Decision Maker (client) and a summary of the purpose of the programming workshop along with the specifics of the first day's agenda. Members of the programming team should be made known along with an explanation of their roles. Likewise it is important to take a few moments for participants to go around the room and introduce themselves and explain what their stake is in the project. Each day after this, new participants will be announced followed by a recap of the prior day and a briefing of the days subjects will be presented. It is best to have enough time for participants to say hello, get a snack, and be seated prior to commencing the first session.

As part of the introductory aspect and for continuous reference, relevant large format photography (cultural aspects, site, artifacts, users, etc.), maps (both macro and micro), and a large format mission statement will be placed in accessible and visible locations easily seen upon arrival (Figure 4.11).

（图4.11）背景照片中展示的是会议参与者发言时紧邻写有项目任务宣言的招贴板。
(Figure 4.11) Screened background photo shows participant adjacent to mission statement poster.

一份清晰的、可调查并容易审阅的讨论记录对于项目的成功非常重要（图4.12）。其中的关键点在于这种记录不能够只是对于讨论简单地描述，而是应该将讨论的内容归纳成想法条目。制作这种类型的文案，参与的专家在写之前应该认真倾听，文案也不会在讨论会中写出，而是会后由有经验的建筑师根据各方意见整理完成。举例来说，几名与会者就在入口处如何欢迎访问者并且同时还能采取相应的安全措施以保证雇员的安全进行的一个10分钟的争论，就可以被归纳为："让公众到达建筑时感到被欢迎非常重要，而与此同时，保障员工的安全也是很关键的一点。"这些记录最好用大号的黑色马克笔写在容易安放的纸板上，一般是A1的尺寸。如果时间允许，可以把这些记录页夹在板子上后在房间里靠边环绕放置。

至少找一次研讨会的机会，将问题卡片分发给与会者，并鼓励他们将自己的想法或者是答案用草图的方式画在卡片上。问题卡片，可以简单地做成75mm × 125mm的大小，上面提前印好问题，那些不太善于在公众场合发言的参与者经常可以利用这样的机会就他们对于例如项目的愿景、项目成功的关键因素、按照优先级排列出的对于项目的希望等问题贡献出非常有想法的答案。一个很重要的方面是要让受访者在答题卡上签上自己的名字。这样做有两方面的原因：一是可以让建筑师将来可以联系到答题者来进一步了解和澄清答题者的本意；二来这样一般就可以避免答题者利用这样的机会发泄个人恩怨。看着很多参与者聚集在一起观看其他人对于一个项目是怎么说的是一件很有趣的事情（图4.13）。

（图4.12）建筑师正在架在支架上的A1尺寸的记录纸上做笔记。
(Figure 4.12) Architect taking typical notes utilizing A1 size flip chart on easel.

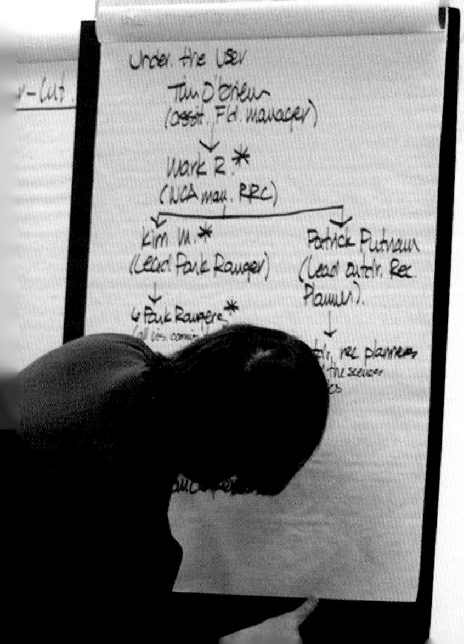

Legible, accessible, and easily reviewable notes of discourse are important to the success of the project (Figure 4.12) Key is the concept that these notes are not transcripts of the discussion, but rather summary statements and ideas. This type of documentation involves expert listening before writing and is not to be learned during a workshop but should be done by an experienced architect. For example, a 10-minute debate involving several participants over how to welcome visitors and maintain security at the entry to protect employees can be summarized as " *it is important that the public must feel welcome while at the same time employee safety is critical*". Notes are best taken with large black ink markers on easily mounted newsprint pads, typically A1size (close to 24"x36"). As time permits, these sheets are ripped from the pad and tacked up around the room.

At least once in each session, question cards are distributed and participants are encouraged to draw sketches of the ideas or answers. Question cards, simple 75mm x 125mm (3" x 5") cardstock, with preprinted questions will often get people reticent to speak in groups to create thoughtful answers to questions regarding their vision for the facility, critical issues for success and to prioritize wishes. An import aspect of using the cards is to ask that respondents sign their names. This is useful for two reasons. First, the architect may need to contact the author for clarification and secondly, this generally prevents personal accusations from being made. It is interesting to see participants gathered around a posting reading what others have to say about a subject (Figure 4.13).

（图4.13）每个研讨会的结尾部分会将问题卡片贴出来供与会者审阅。下方是一张典型的问题卡片。
(Figure 4.13) Participants reviewing question cards that have been posted at the end of each session. Typical card below.

(图4.14)建筑师同项目利益相关者一起绘制空间关系示意图。不同的泡泡尺寸、线型粗细以及颜色可以很明确地标示出功能空间之间的关系。

(Figure 4.14) The architect drawing a space relationship diagram with the stakeholders. Hierarchy of bubble size, lineweight, and color identify relationships between spaces or functions.

使用示意图有助于归纳想法,在这种情况下,建筑师应该在向大家表达自己想法的时候使用示意图将这些想法画下来。当大家可以共同合作绘制出表格的时候,讨论的质量会得到提升。举例来说,当我们讨论记录空间需求的时候,我们可以让参与者将自己的建议填入到安放在支架上的大张的空间需求表格里。一个非常有用的策略(图4.14)是,还是在一张A1尺寸的白纸上,用不同的圆圈代表单独的功能空间、建筑等,然后再在上面用图示的方式记录讨论的内容,诸如空间的进入或离开、各种不同的交通流线(公众、管理人员、员工、服务、消防等)以及空间之间的毗邻关系(积极的和消极的都算)。经常出现的情况是,一个关于建筑的哪一部分是公共的或私密的决定会引发一段很长时间的讨论。举个例子,一个带有很强的志愿者参与功能的政府设施,公共和私密区域的分界线该如何划定呢?

讨论会中的一个关键环节是对于所有参与者都有启迪作用的针对某一相关特定主题的演讲或是研究发现的分享。举个例子,在一个关于蜉蝣的项目中,一个专家可以谈及针对如何保护、处理和储藏纸张的特殊环境要求。

一个值得关注的想法是每天针对公众可以有一个口头的总结性的报告。这将会是研讨会的第五部分,并且一般都会安排在晚上。根据大家感兴趣的程度,也可能会需要一个比较大的空间。

作为建筑师,我们建议至少要举行一次针对公众的会议,以获得对于项目需求更加深入的了解。这种会议可以带来的附加好处是会吸引一些项目的反对者参与到讨论中来。如果可以加入一

In the case that diagrams are useful to summarize thoughts, the architect should draw these as the group expresses their thoughts. The discussion is enhanced when charts are created jointly with the group. For example, when documenting space requirements a large format space requirement form held by an easel can be filled out with suggestions from the participants. One useful technique (Figure 4.14) is to, also on A1 sheets, draw a series of circles (bubbles) representing individual spaces, buildings, etc. and then record input regarding access/egress, various sorts of circulation (public, administration, staff, service, fire, etc.) and adjacencies (both positive and negative). Often, just the determination of what parts of a building are public or private will provoke a lengthy discussion. For example, in the case of a government facility with a strong volunteer component, where is the line drawn?

Key components during workshops are lectures or research findings on a specific subject for the edification of all participants. For instance, in the case of ephemera, an expert may speak about special environmental requirements for the preservation of paper, its handling, and storage.

An idea worth considering is the presentation of daily oral summary reports of proceedings to the general public. This will be a 5th session and occur in the evening, and depending upon interest, may require a large space.

As architects, we recommend at least one public meeting to gain additional insights as to project needs. An additional benefit of this type of meeting is that it will attract project naysayers or detractors. If a political process is involved, this allows the architect to understand who may be against the project and why. In the

些政治性的过程,将会使得建筑师可以了解是谁在反对这个项目以及为什么。在奥利弗牧场科学学校项目中,项目附近布卢戴蒙德的一个镇上的居民强烈反对这个校园的修建,因为他们觉得这个项目将会抢夺他们本已非常有限的水源。当我们展示给他们新建的项目会利用与他们截然不同的水源时,这群人从反对者变成了支持者。

如果一栋建筑的布局是按照不同的部门来划分的,那么安排同部门的总监或是领导进行一对一的沟通就是非常重要的事。这样领导同建筑师之间就可以进行直接的无保留的讨论。但是我们也必须记住,依据功能及环境需求做出符合标准的设计方案是建筑师的工作,我们不需要完全依照个人(领导)先入为主的主观想法来创造出一个空间。设想一个酒店中的餐厅厨房特意为一个身高约为2m或是1.6m的厨师量身定制。因为厨师可能会经常更换,那么这么做可能就是不合适的。但是换个情况,对于一个特定的餐厅可能它的成功完全倚仗某一个厨师,那么所有的设计工作就会围绕着怎么使得这名特定厨师的工作更有效来展开,例如台面的高度是否适合他的使用习惯,搁架板是否放在他容易看到的位置等问题就变得十分重要。

在一个通常要持续长达两个小时的设计任务讨论会中,获取信息方式的多样化十分重要。会议主持人的职责

---

case of the Oliver Ranch Science School, the nearby town of Blue Diamond was strongly against the campus since they felt that it would be competing for their limited water resource. Once it was demonstrated that the school was utilizing water from a different aquifer from theirs, the group changed from detractors to supporters.

When a building is organized into departments or divisions it is important to visit with directors, department heads or other key individuals on a one on one basis. This allows direct, uninhibited, discussion with the architect. It must be remembered, however, that it is the architect's job to design based upon criteria that satisfy specific functional and environmental needs, not necessarily create a space based upon an individual's predilections. Think of a restaurant kitchen in a hotel specifically designed to accommodate a 6'8" or 5'4" chef. Because of the frequency of turnover among chefs this may not be appropriate. On the other hand, an individual chef will be the reason for success of a specific restaurant and so anything that makes him more efficient, like the height of counter tops or shelves is obviously important.

During a programming session, which can last up to two hours, it is important to vary the method of gaining information. It is the

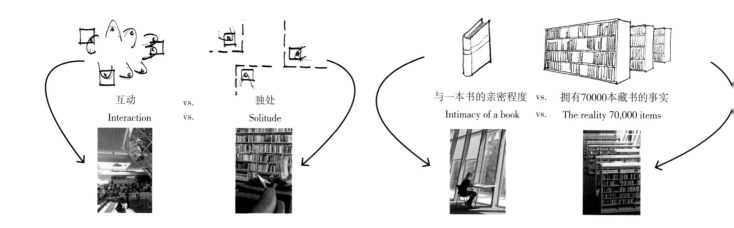

是要引领会议在正确的轨道上进行。关键是要事先准备好针对讨论主题的问题。但实际上不可避免的情况是，会议的参与者经常会偏离议程中计划的主题，转而去讨论一个自己更感兴趣的话题。如果会议的主持人太过教条，死板地将讨论限定在预先设定的范围，与会者有可能会减弱讨论的热情，从而导致没什么人发言。主持人在大家面前以一种同讨论相关联的方式重申需要讨论的主题，或是干脆简单地将讨论的情况（没什么人发表意见）记录在案，都可以让后面的会议回归正轨。

设计任务讨论会的一个很重要的功能就是设定设计目标。这些目标可以集中或是分散到不同的主题中去，例如场地目标、设施目标、教学目标等。当这些目标之间产生冲突的时候，会议主持人必须将这些互相冲突的意见整合为一个有用的目标来描述，或是必须要认识到这个项目客观存在着这些互相矛盾的目标，而由此产生的迷局则需要由建筑师来解决（图4.15）。

session facilitator's (leader's) job to guide the discussion. Pre-prepared questions on topic are essential. Inevitably, participants tend to go away from the agenda topic and want to express a favorite idea or point of view. If the facilitator is too dogmatic about staying on point people tend to loose enthusiasm so noting what one has to say and then either restating it before the group in a way that relates to the discussion or simply including it in the meeting notes will allow the meeting to get back on track.

One of the most important functions of the programming workshop is to establish goals. These goals can be congregated or dispersed among the topics for example, site goals, facility goals, pedagogic goals, etc. When conflicts occur during goal setting the facilitator must be able to meld competing interests into a single useful goal statement or must recognize that there are contradictory goals. The contradictions may simply be elements of the puzzle which the architect must solve (Figure 4.15).

（图4.15）下方的草图展示了设计任务制定阶段所发现的矛盾点，而照片则显示了这些矛盾点最终是怎样影响到项目的设计的。
(Figure 4.15) The sketch diagrams below illustrate the contradictions developed during programming and the photos show how the contradictions influenced the final design of the facility.

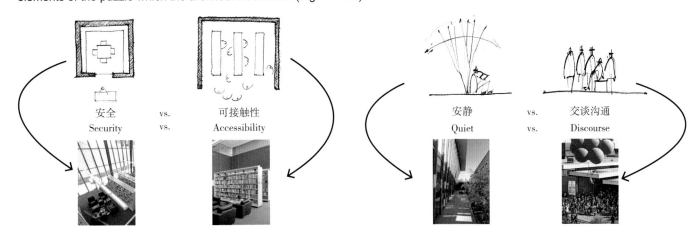

Dialog Checklist   对话清单

Mobilize   动员
- **Schedule**   议程表
  - Meetings and Workshops   会议及研讨会
  - Framework   框架
    - Food   食品
- **Determine Moderators/Facilitators**   确定的会议主持人
  - Talking points   讨论话题
- **Table of Contents**   目录
  - Workshop Agenda   研讨会议程表
- **Equipment Needed**   需要的设备
  - Foam core boards   风波板（KT板）
  - Flip Charts   活动挂图
  - Space Confirmation Sheets   空间确认单
  - Room Sheets   房间列表

Establishing Goals, Needs and Wants   建立起目标和需求
- **Intro/general**   一般性介绍
- **Philosophy**   设计理念
  - Architectural   建筑方面的
  - Design Approach   设计策略
- **Site/context**   场地及文脉
  - Site discussions   场地讨论
    - Site Forces   影响场地因素
    - Access/ egress   进入或离开
    - Utilities   设施
    - Fire   防火
    - Explorations   场地勘测
    - Environmental   环境方面的问题
    - Parking   停车
  - Goals   目标
- **Culture, History, Technology**   文化、历史、科技
  - Goals   目标
- **People**   人员
  - Goals   目标
  - Internal   内部的
    - Leadership   领导
    - Staff   员工
  - External   外部的
    - Visitor analysis   访客分析

- **Facility**
  - General goals and objectives
  - Environmental Goals
  - Parking
  - Operation Schedule
    - Days open
    - Hours, etc.
  - Systems/Utilities
- **Function**
  - Goals
  - General - definition:
    - Function vs. space
  - Function needs
    - Description
    - Diagram
  - Furnishings, Equipment,
  - Space need work sheet
  - Critical issues–identify/goals
  - Functional adjacency diagrams
- **Movement**
  - Service
  - Staff
  - Users
  - Visitors
  - Emergency
- **Space needs**
  - Goals
  - Space list
  - Space need work sheet
    - Functions
    - Critical issues/GOALS
    - Furnishings, Equipment
    - Adjacency
      - *Positive*
      - *Negative*
    - Sensory
    - People
  - Special Discussions
    - Libraries, Hotels
- **Priorities**
- **Summary**

- 设施
  - 一般性的目标和目的
  - 环境方面的目标
  - 停车
  - 运行时间表
    - 开放日
    - 开放时间
  - 设施系统
- 功能
  - 目标
  - 一般性的描述：
    - 功能 vs 空间
  - 需要的功能
    - 描述
    - 示意图
  - 家具、设备
  - 空间需要清单
  - 关键的问题：定义及目标
  - 功能临近关系示意图
- 运动
  - 服务
  - 职员
  - 使用者
  - 访客
  - 紧急情况
- 空间需求
  - 目标
  - 空间清单
  - 空间需求清单
    - 功能
    - 关键问题及目标
    - 家具、设备
    - 临近关系
      - *积极的*
      - *消极的*
    - 感知方面的
    - 人员
  - 特殊问题的讨论
    - 图书馆、酒店
- 优先级
- 总结

文案
Document

文案是对于设计任务信息的文字化的整理汇编。

The document is the written compilation of the program information.

文案是对于设计任务信息文字化的整理汇编。这是设计师的操作手册，其中包含了对于项目利益相关者各种重要的设计标准。建筑师和业主都应该经常回溯到设计任务文案中去提醒他们自己项目的设计目标和实际需求是什么。

我们在这本书中收录了（美国）国家历史路径展示中心项目的实际设计任务文案以方便读者的使用。需要特别注明的非常重要的一点是，设计任务文案是一份"活"的文件，它可以也应该在甲乙双方达成共识的基础上，伴随着建筑师建立起设计概念的同时发展完成。

我们的设想是将这一章节基于实际的文案目录分为10个部分。我们就每一个部分会进行简要地解释，进而会包括这部分实际的文字和示意图。

本书附录中包含了更多的信息，包括表格、日程表，以及由实际的设计任务讨论会产生出来的设计任务文案。这些信息有助于读者进一步理解隐藏在由建筑师主导进行的设计任务制定后面的工作过程。

The document is the written compilation of the program information. It is the designer's handbook and contains all of the criteria important to the stakeholders. The architects and clients should refer back to the program to remind themselves about the project goals and needs.

We have included for our readers use, the actual program document from the National Historic Trails Center. One of the most important things to note is that the Program is a "living" document that can and should by mutual agreement between the client and the architect evolve as architectural concepts are created.

Our approach is to divide this section in to ten parts based upon the actual table of contents. We provide a brief explanation of each part and then include the actual text and diagrams for that section. Section 10.0 Appendix is included because it contains two important sections that could have been included in the body of the text. The first describes the interpretive requirements and the second is additional site information.

The appendix part contains additional information including charts, agendas, and documents that were generated during the actual programming effort. These should be helpful to get an idea of the process behind Architect Directed Programming.

（图5.1）封面应该包含有一张能够清晰概括出项目主题的照片或是图片。对于国家历史路径展示中心项目而言，一张类似于素描效果的黑白对比强烈的照片强调出了移民在大地上留下的痕迹。这些痕迹，是文件中所讲故事的精华，也清晰地反映出了未来建筑需要帮助陈列展示讲述的内容。

(Figure 5.1) The cover should include a photo or graphic that clearly sums up the project. For the National Historic Trails Center, a photograph was sketched in high contrast black and white to highlight the marks on the land left from migration. These marks, essential to the story being told inside, clearly illustrate the story that the building will help tell.

设计方案
国家历史路径展示中心

Design Program
National Historic Trails Center

Line and Space   627 East Speedway   Tucson, AZ 85705   (520) 623-1313

1.0 简介
    1.1 文件描述
    1.2 任务
    1.3 小结

1.1

文件描述部分向读者介绍了设计任务。

1.2

正如书中前面所介绍的，大概的设计任务是已经知道的。在这个案例，即国家历史路径中心的设计任务中，我们发现任务描述的部分需要根据项目实际情况做出适当调整。

1.3

小结部分，通常不会超过2～3页，将文件中的重点用清晰和准确的方式表述出来。这有助于领导者快速准确地传达出项目的精髓。其中的内容包括场地、人员、尺寸、设施的功能、几种不同的预算方案及计划的时间表。

1.0 Introduction
    1.1 Document description
    1.2 Mission
    1.3 Summary

1.1

The document description introduces the reader to the program.

1.2

As explained earlier in this book, the mission is presented so that the overarching purpose of the project is well understood. Here, in the Trail Center Program, it became clear that the mission statement needed to be slightly refined to mesh with the realities of the project.

1.3

The summary, usually no more than 2 or 3 pages describes the salient points of the document in a clear and concise way. This is useful for leadership who want to convey the essence of the project quickly and accurately. Subjects include Site, People, Size and function of the Facility, several budget scenarios and proposed schedule.

2.1

我们认为，在初期制定设计任务的时候，建筑师应该表达出自己的设计理念。这种哲学性的表达可能一般来讲比较空泛和抽象。但很重要的一点是有一件事要表达明确，设计与风格无关，如何去回应场地及当地的历史文化才是启发建筑设计的关键。

2.2

既然设计任务书是一本关于项目信息的手册，那么设计团队就应该可以在里面很方便地查到设计确认所需要的程序及谁是决策人。项目联系人的联系方式也可以在里面方便地查找到。

2.3

在设计任务讨论会结束后，还会有一些特定的信息处于未知的状态，我们将这样的问题称之为未解决问题。未解决的问题最好列出在明显的位置，这样能够促使它们可以被及时解答。另外也可以让其他项目参与者了解这些未解决的问题会怎样影响到项目进度。在一些情况下，有些问题的解答是设计开始的前提，例如"访客的人员构成是什么？"而在另外一些情况下，最终的预算可能要等到方案设计阶段完成后才能够有个答案。

2.0　设计
　　2.1　架构
　　2.2　审查和批准
　　2.3　未解决的问题

2.1

We feel that, early in the program, the architect should express a philosophical statement about his vision for the design. It is meant to be general and abstract. It is important that it is clear that design is not about style but rather how the place and history of the place inspires the architecture.

2.2

Since the program is the handbook of project information, it is convenient for the design team to know what the approval process is and who the approving authorities are. Contact information is conveniently found here.

2.3

After programming workshops are complete, certain information remains unknown we call this the unresolved issues. Unresolved issues are best listed in a prominent location so that they can be answered in a timely way as well as inform participants how these unknowns will affect the project schedule. In some cases the design cannot begin until answers to questions such as "What are the visitor demographics?" are understood in other cases a final budget may itself not be determined until the schematic design is completed.

2.0 Design
　　2.1 Architecture
　　2.2 Review and Approval
　　2.3 Unresolved Issues

3.0 场地
　　3.1 场地描述
　　3.2 地图

3.1

开始的部分我们会有一个在大的文脉背景下的关于场地位置的一般性描述。在这里，我们会指出场地的面积以及与周边道路和交通的关系。高程、景观视线、地质以及植被等问题都会被讨论到。市政设施也会被标注出来。关于场地适建性方面一些政府的要求，例如场地使用性质、退线、建设密度和高度限制等信息都会被记录在案。我们还会归纳总结对于气候的讨论。当然，将主要的场地目标及规范在这里列出也是不错的选择。一个关于场地适建性的清晰而明确的结论会对最终的决策非常有帮助。

3.2

在这份文件的前面已经包含了大量的关于地图以及它们使用方面的讨论。在这里，我们提供了一份详细的位置地图，一起的还有其他的一些专项地图显示诸如地块的进入和离开方式、市政设施、地役权等不同的信息。在这个案例中，景观视线分析示意图是作为一个附录提供的，但在其他一些案例中，它最适合在这个章节出现。

3.0 Site
　　3.1 Description
　　3.2 Maps

3.1

We begin with a general description of the project location within the greater context. Here we indicate site area and relationship to roads and transportation. Elevation, views, geology and vegetation are all discussed. Utilities will be noted. Suitability of the land for building in terms of government requirements such as land use, setbacks, land coverage and height limits is documented. A discussion of climate is summarized. This is a good place to list major site goals and precepts. A clear and concise conclusion regarding suitability of the site is useful.

3.2

Extensive exploration regarding maps and their use are discussed earlier in the document. Here, a detailed location map is provided along with separate maps showing access/egress, a utility map, and critically, easements. In this case, view analysis diagrams have been provided in the appendix in other cases they will be most appropriate in this section.

4.1

当我们讨论访客的问题时，非常重要的一点是我们必须要了解他们的数量及人员构成。显而易见，儿童的需求同成年人非常不一样，而后者的需求又明显不同于老年人。另外一个很关键的点则是我们要了解高峰期的客流量以及在一年中高峰期会出现在什么时间。在这一章节的讨论中还包括顾客会使用什么类型的交通方式。

4.2

这个设施当然是被设计用来支持职员的管理工作。对于现有以及未来的人员框架和组织结构方面的访谈结果都被记录在这里。总结这一结果的一个简便方式就是建立一个人员架构表。在诸如博物馆、访客中心、动物相关设施、医院以及图书馆这样的案例中，志愿者或者是朋友经常会是职员组成中一个重要的组成部分。这些人员虽然不拿工资，但他们的需求也需要充分考虑。

4.3

这个话题涉及了场外支持方面的需求，这些需求包括园艺、食物配送、洗衣、维护和机电设备维修。

4.0 人员
    4.1 访客
    4.2 职员
    4.3 外部支持

4.1

When discussing visitors it is important to understand numbers and specific demographics. Clearly, the needs of children are very different from adults, who in turn have very different requirements from the elderly. It is also critical to have a sense of peak loads and time of the year these will occur. Included in this category is the type of transportation which will be utilized by the patrons.

4.2

The facility will, of course, be designed to support administration and staff. Results of interview findings regarding organization, hierarchy, as well as current and future numbers are recorded here. An easy way to summarize findings is with a staffing chart. Often in the case of museums, visitor centers, animal oriented facilities, hospitals and libraries volunteers or friends groups are an important part of their staffing. These critical people, though unpaid, must be listened to and provided for.

4.3

This topic addresses the need for outside services such as gardening, food service, laundry, maintenance and mechanical servicing.

4.0 People
    4.1 Visitors
    4.2 Staff
    4.3 Off-site support

5.0 房间和空间
　　5.1 设施目标
　　5.2 运营开放时间
　　5.3 服务

5.1

设施目标会在研讨会期间确定，其间会考虑到所有项目利益相关者关于项目的希望和愿景。这包括了关键的针对访客的设计问题、员工的优先级问题以及运营需求问题。

5.2

定义出计划中的运营时间表会引发我们对于环境问题的思考。了解项目在夜间、周末及节假日的运营时间是非常重要的事。

5.3

项目设施可以提供各种不同的服务内容，从简单的信息传达到教育或是为相关的社区活动提供场地。尽管这些内容会随着时间而变化，但尽量地去理解这些服务内容有助于鼓励设计师将设施变得物超所值。

5.0 **Rooms and Spaces**
　　5.1 Facility goals
　　5.2 Operating hours
　　5.3 Services

5.1

Facility goals are determined during workshop sessions and address all aspects of stakeholder's hopes and aspirations for the facility. Included are critical visitor design issues, staff priorities and operational requirements.

5.2

Defining the intended operating schedule informs environmental explorations. Nighttime, weekend and holiday schedules are important to understand.

5.3

Services provided by the facility vary from information to education as well as providing a venue for associated community events. While these will evolve over time based upon the facility itself, getting a sense of these services will encourage the designer to make the facility more than the sum of its parts.

6.1

设计任务书中最重要的章节之一就是关于所有房间及空间需求的小结清单。在这里设计师可以很迅速地看到建筑里的主要空间，它们是如何同其他部分相关联以及它们的实际面积有多大。关键的一点，正如我们在其他部分所讨论过的，空间小结的最后要加全系数以包含一般的交通空间面积。这个系数会根据建筑类型的不同而不同，同时还要考虑是否总建筑面积统计中已经单独计算了电梯、楼梯、卫生间、机电用房以及后勤服务用房的面积。

6.2 ~ 6.8

这一部分的细分章节中单独描述了每一个房间及空间的需要，其中涵盖了大量的细节信息，帮助设计师不仅是理解房间有多大，更重要的，可以了解它的具体功能、可能使用这个空间的人数、一份完整的所需家具及设备的清单、储藏需求、感觉上的需求特别是自然光和景观视线方面的，以及其他一些关键性的问题。这是一个恰当的地方来描述空间之间的邻近关系（参看章节7.1），不同空间之间应该是紧邻或是远离。关键性的问题范围很广，但都遵循这样的原则："如果这一项目或者感觉缺失，无论项目看起来多完美，在使用者看来这个空间都是不成功的。"例子包括对于安静的需求、隐私、充足的面积甚至是某种特殊设备所需要的足够多的电源插座。如果这些关键问题没有解决，那么建筑在这些使用者看来就是失败的。

6.0 房间和空间
   6.1 小结
   6.2 入口或门厅
   6.3 礼品店
   6.4 展览
   6.5 教育
   6.6 办公管理——土地管理局
   6.7 办公管理——基金会
   6.8 维护或支持

6.1

One of the most referred to sections of the program is the summary of all the rooms and spaces. Here the designer can quickly see the building's major spaces, how they relate to each other and their actual floor area. Critically, as discussed elsewhere, the grossing factor is included at the end of the summary to account for general circulation. This varies with the building type and whether such items as elevators and stairs, restrooms and mechanical and other services have been included separately in the gross building area.

6.2 – 6.8

These subsections describe room and space requirements individually and in sufficient detail for the designer to understand not only how large but, importantly, their function, the number of persons involved with the space, a complete list of furnishings and equipment, storage requirements, sensory impacts particularly natural light and views and critical issues. This is an appropriate place to describe adjacencies (see also section 7.1); what spaces should be near or distant from each other space. Also, each of these spaces or areas is an ideal place to discuss Critical Issues. Critical Issues are diverse, widely ranging and address the statement that: "If this item or feeling is omitted, no matter how seemingly trivial, the space will not be successful in the opinion of the user" Examples include need for quiet, privacy, sufficient floor area or even adequate power outlets for special equipment. Remember: If the critical issue for a user is not met, the building will be a failure in that person's opinion.

6.0 Rooms and Spaces
   6.1 Summary
   6.2 Entry/lobby
   6.3 Gift Shop
   6.4 Exhibits
   6.5 Education
   6.6 Admin. - BLM
   6.7 Admin. - Foundation
   6.8 Maint./support

7.0 空间关系
　　7.1 空间邻近倾向性列表
　　7.2 相关感知需求列表

7.1

空间邻近倾向性列表是在一张图表中将章节6.0中描述的所有空间都列在互相垂直的两条轴线上。于是这些空间相交的部分就可以用数字来表示它们之间的关系。数字"1"表示两个空间之间有着很强的关系，换句话说，例如总监办公室和秘书的办公室这样的空间必须要紧挨在一起。数字"2"表示空间之间希望有联系，两个副总监的办公室应该彼此靠近，但并不是必须要紧邻。如果使用数字"3"，则表示空间之间的关系是中性的，换句话说，设计师可以根据需要来决定空间的位置安排。数字"4"表示两个空间不希望挨得太近，例如打印区一般不会和会议室靠得太近，因为前者会产生人流和噪音。如果数字"5"出现在了两个垂直数列的相交处，例如总监的办公室和员工休息室，那么就意味着这两个空间非常不希望靠得太近，建筑师应该确保它们之间保持一定的距离。正如你可以看到的，章节6.0里的空间关系表格中所总结出的内容对于建筑师来说是一个可以很方便使用的工具，在这个基础上，就可以开始构建建筑的平面和剖面了。

7.2

感知需求列表描述了与每一个空间相关的不同的感知方面的可能性，例如视觉上的（景观视线、眩光以及灯光照明），或是热工方面的（不同的热传导方式）。这个列表如果可以将这些需求的重要性用数字标出则会对设计师非常有帮助。

7.0 Relationships
　　7.1 Preference Matrix
　　7.2 Sensory Matrix

7.1

The Adjacency or Preference matrix is a chart which lists all spaces described in Chapter 6.0 on each of two perpendicular axes. These spaces are then related to each other by a numbered key. The number "1" indicates a strong relationship between two spaces; in other words spaces such as the Director's office and the secretary's office must be next to each other for optimum success. "2" means a relationship is desirable; two vice presidents should have their offices nearby but it is not crucial. If a "3" is used, the relationship is neutral, in other words, it is the designer's choice where to position this space. The number "4" is used to communicate that it is undesirable for two spaces to be adjacent; i.e. a copy room is better not being placed next to a conference room because of traffic and noise. If the number "5" appears in the line joining two areas such as the director's office and the break room the relationship is highly undesirable and the architect should make sure that they are not near each other. As you can see this chart summarizes relationships described in Section 6.0 and is a very convenient tool for the architect to utilize as he starts diagramming space relationships fundamental to creating building plans and sections.

7.2

The Sensory Matrix describes each space relative to various sensory possibilities such as visual (views, glare and lighting), and thermal (various sorts of transitions). This matrix is most useful to the designer when keyed to show how important these requirements are.

8.0 预算
　8.1 一般性的
　8.2 不同条件下的

8.1

当讨论预算的时候，通常对于决策人最重要的事情就是在进行一般性的建造成本比较时将未来可能的意外及通货膨胀等因素一并考虑。当通货膨胀率比较低的时候，例如每年2%，很关键的一点就是在进行预测的时候要将一个系数加入到计算中。正如你能够看到的，从最开始制定预算到施工过程的中期有五年的时间，预算要增加额外的10%才能满足要求。如果忽视这个增量就会有麻烦。在一种更坏的条件下，例如我们在内华达州拉斯维加斯附近的项目所经历的，我们正好经历了一个建造的高峰期，通货膨胀率是每个月1.5%，年通货膨胀率达到了20%。经过了从概念设计到施工中期的4~5年时间，项目预算增加了一倍。

8.2

在设计任务制定阶段最好可以向客户提供两到三种不同条件下的预算方案。在下面的例子中你将会看到预算A、B和C。这些方案向客户展示了除了严格按照设计任务完成整个项目的建造，还可以有一些别的策略，例如通过将项目分周期建造来缩减成本。

8.0 Budget
　8.1 General
　8.2 Scenarios

8.1

When discussing the budget, often the most critical issue for the DM, it is best to commence with a general discussion of comparable building costs along with other factors such as contingencies and inflation projections. When inflation is low, say, on the order of 2% annually, it is critical to allow a congregate number into your projections. As you can see, if it is five years from the time of initial budgeting to the mid-point of construction an additional 10% will be needed to meet requirements. Omit this number and your building will suffer. In a worst-case scenario, such as we were affected by on our projects near Las Vegas, Nevada, at the height of a building boom, inflation was 1.5% per month or on an annualized basis around 20%. Over four or five years from concept to construction mid-point the project budget will double.

8.2

The program is a good place to present the client with two or three budget scenarios. In the example following you will see budgets A, B and C. These explain costs to the client to build both the complete project as programmed and various other strategies for reducing cost such as phasing of construction.

9.0 日程表
9.1 日程表

9.1

通常对于客户来说,相比预算第二重要的事情是日程表。客户想知道什么时间建筑可以交付使用,这样可以提前计划长期的预算、员工培训以及开业庆典等一系列事宜。在设计任务书中所展示的时间表需要将主要的项目节点列出并表明预计的完成时间。这里列出的只是一个大致的时间表,要求太多的细节信息还为时尚早。在前面"开始"的章节有关于时间表方面的介绍。

9.0 Schedule
9.1 Schedule

9.1

Often, for the client, second in importance to budget is the schedule. The client wants to know when the building will be ready to occupy so that long lead budgeting, staffing and opening festivities may be planned. The schedule shown in the program should list major milestones along with expected completion dates. Here it is best to be quite general as it is too early for much detail. An introduction to scheduling is in the chapter **Beginning**.

10.0 附录
10.1 附录

10.1

第一个附录是来自于展览设计公司"Hilferty and Associates"的关于展示意图的报告。这份报告包含了关键的访客流线和展示内容之间关系的示意图。

重要的景观视线示意图也放在了附录里,以免设计任务书文件的主体过于冗长。这些内容一般会是章节3.0中的一部分。

10.0 Appendix
10.1 Appendix

10.1

The first Appendix item is the report of the interpretive intent of the exhibit designers, Hilferty and Associates. This report includes the critical Visitor Flow and Content Relationship Diagram.

Important View Diagrams were included in the Appendix to reduce the volume of the main body of the program. These normally would be part of section 3.0.

附录
Appendix

在制作设计任务文件时采用的表单实例（带有原始的标注）。
Examples of actual worksheets (with original annotation) used in developing the document.

讨论会流程图

这张流程图是在国家历史路径中心项目规划初期阶段制定的。正如你可以从上面的日期中看到，这张表格是在和项目决策人会谈之后绘制的，以期获得批准并将项目向前推进。这张表格上显示了包括所有重要项目节点的一个完整的网络示意图，上面展示了每一项内容的起始和结束时间。在表格中包含了所有在概念设计汇报前需要完成的工作，给建筑师提供了一个可以参考并且根据需要可以随时更新的宽泛的工作时间表。示意图中大量的留白是方便随着项目的进程有实际状况发生时，建筑师可以手写标注在图表上。

A.1 **workshop flow chart**
This Flow Chart was developed during the initial stages of planning for the National Historic Trails Center. As you can see from the dates, the chart was developed after meeting with the DM and gaining final approval to proceed. The chart shows a complete network diagram with milestones that include start and end dates. Each activity that needs to be accomplished before presenting conceptual design is included, providing the architect a broad schedule that should be referred to often and revised and updated as needed. Ample margins and white space within the diagram allows handwritten notes to be used as the project progresses and actual conditions effect the schedule.

讨论会日程表

设计任务制定研讨会时间表展示了线和空间事务所是如何在国家历史路径中心项目前期研讨会的组织过程中收集信息的。

A.2 **workshop agenda**
The programming workshop agenda on the following pages illustrates how Line and Space organized the process of gathering information during the National Historic Trails Center project workshops.

讨论会概述

此概述是实际案例中会议主持人用来指导组织会议的。正如你们能看到的那样，对于会议主持人来讲，时间的管理是关键，他要能够在有限的时间里引领与会者并激发有价值的对话。

A.3 **overview of workshop**
The outlines on the following pages show the actual documents that facilitatorsused to direct programming sessions.As you can see, time management is key as the facilitator must lead the group and generate valuable conversation and dialog.

问题卡

让项目的利益相关者们就项目的一些关键问题在问题卡上简要地填上答案是一种收集人群意见的简单方式。

A.4 **question card**
Having stakeholders fill in Question Cards that ask for concise answers about key aspects of the project is an easy way to gather consensus amongst a group.

研讨会上使用的空间问题讨论图页

这是一张典型的空间清单，一旦建筑师同项目利益相关者讨论并确定了每一处功能空间，他就可以使用相关信息例如空间的容量、家具及设备配置来制定出项目大致所需要的面积。

A.5 **workshop space sheet**
Atypical space sheet.Once each space is discussed and finalized with stakeholders,the architect can use information such as needed capacity,furnishings and equipment to generate a rough square footage for the project.

词汇表
Glossary

An **Agenda** is simply a list of things to be accomplished. In this case we refer to a written document focused upon topics for a single meeting.

An **Arroyo** also called a wash, is a dry creek, stream bed or gulch that temporarily or seasonally fills and flows after sufficient rain.

A **Confirming Authority** is a person or group that the DM reports to or has final project approval (i.e. Corporate Board, City Council, School Board, or, Federal Entity). For an example of a Federal Entity hierarchy of Higher Authority, Google search for "BLM Manual Bureau of Land Management MS 1201" and see the document titled: *US Department of Interior Bureau of Land Management, 1201- Organization.*

The **Contingency** is a number, usually expressed as a percentage, that helps to mitigate the uncertainty of that number. When a project starts, the area required for a building (or budget) is highly conceptual and the contingency can be a huge factor. i.e. 200% or more. As knowledge increases and the building use/structure is defined in loose terms, the contingency may be 50%. At the end of programming, the contingency can be 15 or 20% to account for both area and budget doubts. By the time the detailed drawings are completed and detailed estimates are prepared, uncertainty regarding the building is reduced to a minimum, the contingency may be reduced to 5%-10%. Construction start dates, construction duration, inflation, interest rates at a future date and the ability to finance the project at a rate certain are all important considerations.

日程表是指将需要完成的事情列成清单。在这个案例中，我们是特指将一次会议中需要讨论的议题用文案的方式集中记录下来。

河床也可以叫作河道，是一条干涸的溪流，平时没有水，但会临时或是季节性的在充沛降雨过后形成湍急的水流。

权威确认人是指一个人或是团体，他或他们拥有最终的项目确认权，或是项目决策人要向他们报告（例如企业的董事会、城市议会、学校董事会或是联邦实体等）。拿一个上一级的联邦实体为例，用谷歌搜索"BLM 手册，土地管理局总体规划1201"，会看到文件上的抬头是《美国内政部土地管理局，1201- 组织规划》。

偶然性系数是一个数字，它通常被表达成一种百分比，帮助减轻数字当中的不确定性。当一个项目开始的时候，建筑所需要的面积（或是预算）是非常概念性的，这时候的偶然性系数可能会非常高，例如200%或是更多。当关于项目的已知信息增多，并且建筑的用途或结构形式被大致限定后，这个系数可能会降至50%。在设计任务制定接近尾声时，考虑到对于面积及预算方面的疑问，我们在计算中还会增加15%~20%的偶然性系数。在详细的设计图纸完成后，一份详细的预算会准备好，这时候不确定的因素已经被降到了最低，偶然性系数可以被进一步缩减到5%~10%。之所以还会有这个系数是因为施工开始的时间、施工持续的时长、通货膨胀、汇率的变化以及项目未来融资的情况都是必须要被考虑的因素。

Contract Documents include contracts between the owner and contractor, bonds, insurance and general contractual conditions relating to the building project. Also included are the architectural drawings, specifications, addenda, and any modifications of the contract i.e. change orders.

Critical Path Method (CPM) a diagram which links contingent activities and their durations in a logical network leading to the completion of the work.

The **Decision Maker (DM)** is the person who has the authority to approve the project design (possibly subject to a higher authority); this may be an Organization President, a Museum Director, a School Superintendent or Principal, a University Facility Project Manager, etc…

A **Determinant** is a factor that decisively affects the nature or outcome of something.

Dialog refers to stakeholders discussing needs and wants with the architect as well as exchange of ideas leading to the formulation of environmental and building goals.

Dogma is a settled or established opinion, belief, or principle.

A **Facilitator** is someone who helps a group of people understand their common objectives and assists them to plan how to achieve these objectives; in doing so, the facilitator remains "neutral" meaning he/she does not take a particular position in the discussion.

A **Flow Chart is** a graphic representation, using symbols interconnected with lines, of the successive steps in a procedure or system.

合同文件包括甲乙双方签订的合同，以及其他一切与合同条款相关联的文件例如债券、保险或是合同条款的前提约定。这也包括了建筑图纸、设计说明、项目日程表以及所有相关的设计变更单等。

关键路径方法(CPM)是一张项目进度表，它将项目开发过程中的每一个关键动作用一个符合项目运行逻辑的网络图组织在一起并标注上该动作需要花费的时间。

项目决策人(DM)是一个有权力确认设计方案的人（也有可能是更上一级的机构）；这个人可能是一个组织的主席、一个博物馆的总监、一个学校的总裁或是校长、一个学校基建处负责人等。

决定因素是指会直接影响到事物本质或是产出结果的一个因素。

对话指的是项目利益相关者同建筑师之间讨论需求以及相互之间交流有关如何达到环境及项目目标方面的看法。

信条是一个已经达成共识的意见、信仰或是原则。

会议主持人是一个可以帮助大家了解他们共同的目标并协助他们计划怎样来达到目标的人。为了达到这一目的，会议主持人必须保持态度上的中立，在会议的讨论中不持立场。

流程图是一种平面图形表示方法，利用符号之间连线来表示程序或是系统的过程步骤。

Geography refers to all aspects related to the land. This includes terrain, geology, biology, zoology, climate, population and land use.

A **Grossing Factor** is a number which adds space for circulation, structure, exterior walls, interior partitions, HVAC equipment. The grossing factor varies according to building and space type and whether support spaces such as storage, mechanical rooms, etc. have been included in the space allocation. In our experience, this number averages between 20% - 35% for low structures and 30% - 40% for high rise buildings.

HVAC Heating, Ventilation, and Air Conditioning, the technology of indoor environmental quality.

Long-Lead items are things or processes which can take a long period of time to accomplish or acquire. Examples include visitor analysis studies, environmental studies and impact reports, or certain construction processes or equipment. These items/processes should be scheduled early enough that their availability or completion is not a factor in delaying the end result.

A **Node** is a key public gathering place that encourages people to linger and socialize.

An **Organization Chart** (often called organizational chart, or org chart) is a diagram that shows the structure of an organization and the relationships and relative ranks of its parts and positions/jobs.

Precepts are guiding commandments or directions given as a rule of action or conduct, i.e. "form follows function".

Programming is the process of discovering and documenting the essential elements of a project.

地理指的是一切同土地相关的知识。它包括地形、地质、生物、动物、气候、人口构成及土地利用。

加权系数是一个数字，它使得最终的面积计算中可以考虑到交通、结构、室外墙体、空调水暖等设备所需要的面积。这个数字会根据建筑及空间类型的不同而有所变化，同时还要考虑到面积分配时是否包括了一些辅助空间，例如储藏、机电设备用房的面积。根据我们的经验，底层建筑的加权系数平均在20%～35%，而高层建筑则要升至30%～40%。

HVAC 取暖、通风和空调设备，提升室内空气质量方面的技术设施。

长期项目是指那些需要经过比较长的时间才能完成或是获得的事情或是过程。这样的例子包括访客分析研究、环境研究及影响报告，或者是特定的施工过程或是设备。这样的项目或过程需要提早规划以避免造成对于整个项目进程的延误。

一个景观节点是一个关键的公众聚集空间，它用来鼓励人员在这里逗留和社交。

一个组织架构图是一张示意图，它上面标明了机构的组织结构以及人员之间的关系和他们的职位及工作职责。

准则是指指导性的意见、对于行动或是行为规范来说引领性的原则，例如："形式追随功能。"

设计任务制定指的是发现并记录项目关键元素的过程。

A **Schedule**, in this context, is an organizational tool for accomplishing certain activities within a specific time frame. The following paragraph was reproduced from Jelen's Cost and Optimization Engineering, author F. Jelen, copyright 1983, ISBN 0-07-053646-5 and has been reproduced on the basis that is for nonprofit educational purposes.

- Level 0: This is the total project and is, in effect, a single bar spanning the time from start to finish.

- Level 1: This schedules the project by its major components. For example, a level 1 schedule for a process plant may be broken into process area, storage and handling area, site and services, and utilities. It is shown in bar chart format.

- Level 2: Each of the level 1 components is further subdivided. For example, utility systems are broken into water, electrical, gas, sanitary, etc. In most cases, this schedule level can only be shown in bar chart format although a bar chart with key constraints may be possible.

- Level 3: The subdivision continues. This is probably the first level that a meaningful critical path network can be drawn. It is also a good level for the project's overall control schedule since it is neither too summarized nor too detailed.

- Levels 4–?: The subdivision continues to whatever level of detail is needed by the user. When operating at these more detailed levels, the planners generally work with less than the total schedule. In most cases these "look-ahead" schedules span periods of 30–180 days. The user may utilize either bar chart or CPM format for these schedule.

日程表，在这本书中，是一种为了在特定的时间段完成特定的任务所使用的时间组织工具。下面的段落摘选自著作《成本和优化工程》，作者是F·Jelen，1983年出版（ISBN 0-07-053646-5），并且在以为非营利性教育目的服务的基础上被转载。

层级0：这是整个项目，实际上是一条单独的横条覆盖了从项目开始直至结束的整个过程。

层级1：这部分日程安排将项目分成了几个主要的组成部分。举例来说，一个加工厂项目的时间表可以被细分为加工区、仓储、装卸区、项目场地、后勤服务以及设备设施等多个方面，并被呈现在一张柱状表格里。

层级2：每一个层级1中的分项都可以进一步细分。举例来说，设备系统可以被细分为水、电、气、卫生等。在大多数情况下，这个层级的时间表只能用柱状表格来表达，当然，可能会带有一些限制条件。

层级3：进一步细分。这可能是第一个可以画出有意义的关键路径网络图的层级。这个层级也很合适来制作出一个项目的总体控制日程表，因为它既不是太笼统，也不是过分详细。

层级4-？：细分的过程可以一直继续下去，直到用户需要的详细程度为止。当操作这些更细节的层级时，规划人一般花费的时间相比于制定项目整体时间表时更少。在大多数情况下，这些"前瞻性"的时间表的跨度一般是30～180天。使用者可以利用柱状图表或是关键路径网络图来绘制这些时间表。

Site Analysis looks at the facts (Site Documentation) with an eye to explaining how they will affect the design.

Site Documentation is an inventory that records the facts having to do with the site, generally summarized in final form, on a single map.

Stakeholders are those individuals that can affect or be affected by an organization's actions, objectives and policies. Examples include directors, employees, government (and its agencies), owners (shareholders), and the community.

Typology is a classification according to general type i.e. residential, commercial, retail, industrial, mixed-use.

Venue describes the place, in regard to programming, where discussions will take place.

Workshops are meetings where a group of people gather to attempt to generate ideas and education themselves about particular subjects.

场地分析深入观察所有与场地有关的事实（场地资料）并解释其会怎样影响到未来的设计。

场地资料是一个记录了所有同场地有关系的事实的清单，一般来说最后这些信息都会被归纳记录到一张地图上。

利益相关者是那些会影响到机构的行动、目标和政策或是会被这些因素影响到的个体。这样的例子包括总监、雇员、政府（以及它的代理机构）、业主（产业持有者）和社区。

项目类型是指按照一般的使用功能进行的分类，例如居住、零售、工业、混合功能等。

会场描述的是一个地点，针对设计任务讨论这个过程，则是指会议讨论所占用的场地。

研讨会是指一群人聚集在一起，就一个特定的话题进行讨论以期互相学习并产生一些想法。

关于作者
About the Author

莱斯·沃里克先生同朱建平先生在一起讨论设计方案。
Les Wallach collaborating with Zhu Jianping.

关于作者

莱斯·沃里克先生是美国建筑师协会院士团成员，他在1978年于亚利桑那州的图森市创立了线和空间建筑师事务所。

莱斯先生在设计任务制定、场地规划以及绿色建筑设计方面有着超过35年的经验。作为线和空间事务所的创意总监，他参与到了设计工作的各个方面。他的设计已经获得了超过100个设计奖项并在全世界范围发表。

About the Author

Les Wallach, FAIA, a member of the American Institute of Architects College of Fellows, founded Line and Space, the Tucson, Arizona architecture design firm, in 1978.

Les has over 35 years of experience programming, master planning, and designing resource conserving projects. As Creative Director for Line and Space he is involved in all aspects of design. His work has received over 100 awards and has been published worldwide.

图书在版编目（CIP）数据

信息收集-形式背后的逻辑　设计前奏：汉英对照/［美］沃里克编著；金雷译. —北京：中国建筑工业出版社，2016.3

ISBN 978-7-112-19227-4

Ⅰ.①信… Ⅱ.①沃… ②金… Ⅲ.①建筑设计-理论-汉、英 Ⅳ.①TU201

中国版本图书馆CIP数据核字（2016）第050064号

责任编辑：唐　旭　杨　晓
责任校对：陈晶晶　李美娜

信息收集-形式背后的逻辑　设计前奏
［美］莱斯·沃里克　编著
金雷　译

*

中国建筑工业出版社出版、发行（北京西郊百万庄）
各地新华书店、建筑书店经销
北京锋尚制版有限公司制版
北京盛通印刷股份有限公司印刷

*

开本：889×1194毫米　1/20　印张：7⅖　字数：238千字
2016年7月第一版　　2016年7月第一次印刷
定价：68.00元（含光盘）
ISBN 978-7-112-19227-4
（28468）

版权所有　翻印必究
如有印装质量问题，可寄本社退换
（邮政编码 100037）